THE PATHWAY TO KNOWLEDGE

ROBERT RECORDE

THE PATHWAY TO
KNOWLEDGE

A FACSIMILE OF THE
FIRST EDITION
IMPRINTED AT LONDON BY
REYNOLD WOLFE
1551

RENASCENT BOOKS

THE PATHWAY TO KNOWLEDGE

First imprinted by Reynold Wolf in 1551
Subsequent Editions
Imprinted by I. Kingston for Ihon Harrison
1574
Imprinted by Iohn Harison for Iohn Harison
1602

This facsimile edition published by
TGR Renascent Books
27 Springdale Court
Mickleover, Derby DE3 9SW
United Kingdom
2009

Paperback edition first published 2013

ISBN 978-1-4820823-8-8
www.renascentbooks.co.uk

Printed and bound by
CreateSpace, Charleston, South Carolina, U.S.A.

for
STANLEY

INTRODUCTION

This book is a facsimile of the first edition of Robert Recorde's *The Pathway to Knowledge*, originally printed in London by Reynold Wolfe in 1551. Recorde was a typical Renaissance man, a physician at the courts of Henry VIII, Edward VI and Mary Tudor, a mathematician and one of the outstanding scholars in England in the middle of the sixteenth century. This book is the second in a series of justly famed mathematical works with the following titles: *The Grounde of Artes, The Pathway to Knowledge, The Gate of Knowledge, The Castle of Knowledge, The Treasure of Knowledge* and *The Whetstone of Witte*. Of these six, *The Gate* and *The Treasure* are no longer extant.

The second book in the series (the one you now hold) deals with geometry; the other books cover arithmetic, astronomy and algebra. It is probable that everyone who studied some aspect of mathematics in Elizabethan England would have done so by joining Recorde for at least some of the way along the metaphorical journey suggested by the ordered progression of his titles. Many readers, in an age of great voyages of discovery and enterprising seamen, would no doubt have been young men with aspirations to serve as navigators or gunners aboard the country's ships of war and trade – positions requiring no small amount of mathematical skill.

This book is the only one of Recorde's mathematical works not written in the form of a dialogue between a master and a scholar. It is the earliest geometrical work in English and it was in general use until the middle of the seventeenth century as an elementary textbook. It was reprinted again in 1574 and finally in 1602. The *Pathway* was originally planned to consist of four books. This volume contains only the first two. On the eighth page of the epistle to King Edward, Recorde says that he will 'put forth the other two books, which should have been set forth with these two, if misfortune had not hindered it'. On the third page of the preface to the second book, he repeats that he will 'set forth the other books which are now left un-printed, by occasion not so much of the charges in cutting of the figures, as for other hindrances, which I trust hereafter shall be remedied.' It seems, therefore, that books three and four were actually completed, although they were never published at any later date. If *The Pathway* is compared with Euclid's *Geometry*, it will be seen that the problems have been segregated from the theorems, with Book I containing almost all of the problems in Books I, II, III and IV of

Introduction

Euclid and in virtually the same order. Likewise, Recorde's Book II has all of the theorems of Euclid, Books I, II and III. According to Johnson and Larkey, Robert Recorde was the first modern writer to divide Euclid's propositions into two classifications, that is, 'problems' which demanded that something be done, such as constructing a triangle, given two sides and the included angle, or inscribing a square in a given circle; and 'theorems' which require the general proof of some geometrical truth. Ramus has usually been credited by his biographers with being the first to suggest this distinction, which makes for greater clarity in the teaching of geometry. However, the earliest work by Ramus which contains this classification is his *Prooemium Mathematicum*, published in 1567, sixteen years after Recorde's *Pathway*. Recorde assembles the definitions at the beginning of Book I, while the 'grantable requests' (postulates) and 'common sentences' (axioms) are in Book II. It is interesting to note that his definitions are much simpler and more practical than those of Euclid. For an example, see his definition of a 'point' and compare with Euclid.

BRIEF BIOGRAPHY

Robert Recorde was born sometime between 1510 and 1512 in Tenby, Pembrokeshire. He was the second son of Thomas Recorde and Rose Jones. Practically nothing is known of his childhood and the first thing in his life about which we can be certain is his entrance into Oxford University in about 1525. It is not known what he studied but we may assume it was the usual (for the time) *trivium* of grammar, logic and rhetoric, followed by the *quadrivium* of arithmetic, geometry, music and astronomy. He graduated with a B.A. in 1531 and was elected a Fellow of All Souls College in the same year. He may have taught at Oxford for a few years but the evidence for this is scanty. At some time he moved from Oxford to Cambridge, where he studied for an M.D. and graduated in 1545 at the age of 35. He then moved to London, where for a few years he practised medicine. In later years he was always to describe himself as 'physician'. A defining moment in his life occurred in 1549 when he was appointed Controller of the Bristol Mint. It was during his time there that he made a very powerful and ruthless enemy. Sir William Herbert was sent by Edward VI to help suppress a revolt by John Dudley, Earl of Warwick, in the west country. Herbert demanded that Recorde divert funds from the mint to pay and support his army, but Recorde refused on the grounds that the order did not come from the king. Her-

Introduction

bert countered and accused Recorde of treason. He was lucky to incur the mild penalty of confinement to court for 60 days.

However, apparently all was later forgiven because in 1551 he was appointed general surveyor of Mines and Monies in Ireland. He was placed in charge of the Wexford silver mines and also became the technical supervisor of the Dublin mint. In the meantime, Sir William Herbert was created Earl of Pembroke for his services to the crown during the rebellion, and there was continued animosity between him and Recorde. Although the silver mines at Wexford had great potential, the enterprise was largely unsuccessful, mainly due to a lack of royal investment and the imperfect state of mining technology. The mines closed in 1553 and Recorde was recalled to England. Upon the accession to the throne of Mary, the daughter of Henry VIII, Recorde's old enemy the Earl of Pembroke was made a privy councillor for his support of Mary's claim to the throne. For some strange reason, Recorde chose the moment when Pembroke was strongest to try and get his revenge, charging him with misconduct in gaining his court positions. The allegation was probably true, but Pembroke was in favour with the monarchy and so had almost perfect immunity. He responded by suing Recorde for libel.

There was a hearing in January 1557 and Recorde was ordered to pay the huge sum of £1000 compensation. He either could not or would not pay and so was sentenced to imprisonment in the King's Bench Prison in Southwark, for debt. Whilst in prison he made his will, leaving small sums of money to various people, including £20 to his mother. The date of his death is not known with any certainty, but is generally supposed to have been in the later part of 1558, only a short time after making his will.

The following notes are provided as a guide to reading, understanding and enjoying this facsimile edition of Recorde's remarkable book on geometry.

PAGINATION

The pages are not numbered individually as in modern practice. However the signatures are identified and some page order is imposed by the consecutive numbering of the various conclusions and theorems. In a few places the sixteenth century compositors have confused this numbering – for example, two consecutive conclusions are both numbered 37 (instead of 37 and 38), or again, theorem 26 follows theorem 24 with the omission of 25.

Introduction

However, for the benefit of modern day readers who expect consistent pagination, modern page numbering is applied over a line drawn above Recorde's text.

SPELLINGS AND PUNCTUATION

Dictionaries and standard spellings did not exist when *The Pathway to Knowledge* was first printed and many words are not even spelt phonetically. Therefore many spellings in the book appear peculiar to the modern reader, but a little practice at reading Early Modern English soon renders the text intelligible. Many familiar words look strange simply because, unlike modern spellings, they end with the silent letter e and the last consonant might or might not be doubled, hence mane or manne (man), and rune or runne (run). The letter y is often used in place of i, for example fynde (find) or fyrste (first). Note that early printing conventions were to use the terminal letter s at the end of words, as today, but the long form everywhere else, for example poſſeſs (possess). The letters u and v were not considered to be two distinct letters, but different forms of the same letter. Typographically, v was often used at the start of words and u elsewhere, hence vnmoued (unmoved) or vnloued (unloved). But conversely, the letter v was often used where today we would expect the letter u, as in, for example, thervnto (thereunto). Neither were the letters i and j considered distinct, so that the name John would appear spelt as Iohn. In short, expect to read words such as sum, divisor and just for example, as ſumme, diuiſor and iust.

CONTRACTIONS

On the majority of pages the words 'the' and 'that' are contracted to yᵉ and yᵗ respectively, with the small letters e and t placed directly above the y. The word 'with' is often contracted to wᵗ and similarly 'which' is sometimes contracted to wᶜʰ, again with the t or ch in smaller point directly above the w. Of course, all these words should be read with their full pronunciation.

DIACRITICAL MARKS

Diacritical marks have been used to abbreviate printed words ever since Gutenberg and early English printers adopted the same conventions that Gutenberg did for Latin texts (which he copied, in turn, from the handwritten texts of medieval scribes). Diacritical marks are used on almost every page of *The Pathway to Knowledge* to indicate the omission of the consonant m or n where this follows a vowel. The

Introduction

missing letter is indicated by placing the mark (a bar) over the vowel. Instances are exāple (example), quotiēt (quotient), ī (in), nōbre for nombre (number) and chaūce for chaunce (chance). All such abbreviated words with diacritical marks should be read of course with their full pronunciation.

TYPOGRAPHICAL FEATURES

All the contractions and abbreviations found in the pages of *The Pathway to Knowledge* are compositors tricks to help in the justification of entire paragraphs – something that was considerably easier in the days before standard spellings and orthography. Justification of paragraphs was not then merely a cosmetic feature (as it is today). Early printers would be laying out movable metal type into a square wooden frame and if the frame was not completely filled, the types might move under the action of the press and smudge the ink. In other words, each line of each paragraph had to extend fully from left to right, or the page would be unprintable. One way for Renaissance printers to do this might have been by inserting blank spaces of suitable lengths between the words of each line, but this is not a satisfactory solution. The result is usually 'rivers' of blank space flowing down the page, which seriously interrupt reading and which was recognised as a problem from the very earliest days of printing. Hence the use of aggressive hyphenation, contracted words, diacritical marks and variant spellings like hed (which has three letters), head (which has four), or hedde (which has five), all very useful when striving to obtain justification and spelling is not a problem. The compositor would use any or all of these tricks at will in order to obtain a solid block of text on each page.

FAULTS

There are no errata listed in *The Pathway to Knowledge*. However, the perceptive reader will find a number of faults and errors and, of course, no attempt has been made to indicate or remedy these in this facsimile reprint. In a number of cases the running head at the top of each page is not consistent in reflecting the content of the page beneath it. It is not unusual to find letters upside down, so that the letter n looks like u and m looks like w (and vice-versa). Similarly, letters are sometimes juxtaposed so that 'conclusions' for example, is printed as 'conclvsiosns'. More seriously, a number of geometrical drawings are mis-labelled or have points, sides or vertices not labelled at all, so that explanations in the text which refer to the figures are not always as clear as the author intended. Almost certainly this is due to mistakes made when the figures were originally engraved and which were not subsequently noticed by the printer, or it may simply have been too expensive to re-cut them. When readers encounter these sorts of typographical errors in the book, remember that they occur in the original printing and are faithfully reproduced in this facsimile reprint.

THE TITLE

On the cover of this book, the title is given as *The Pathway to Knowledge*, that is, with modern spellings. However, a reference to the title page of the facsimile shows the title as *The Pathway to Knowledg*, the last word having no final e. The 1574 edition had variant spellings and the title was rendered as *The Pathwaie to Knowledge*. The 1602 edition had a further peculiarity, where the w in *Pathway* was formed by duplicating the letter v, so that it appeared as *The Pathvvay to Knowledge* (this was not an unusual convention in printed works at this date – but note here that the word *Knowledge* does have a normal w). Conventional spellings are the least confusing to the modern reader, and are therefore judged to be the most suitable to appear on the cover.

THE EPISTLE

Robert Recorde uses the epistle to address King Edward VI, the son of Henry VIII. Known to history as the 'child king', Edward died just before his sixteenth birthday, in 1553. Acquiring a reputation for scholarship during his all too brief life, a short biography by Hoak relates that Edward's '...much used copy of Robert Recorde's *The Pathway to Knowledge*, a popular text on geometry, and an edition of *Euclid*, clearly point towards instruction in mathematics and astronomy. Edward's astronomical brass quadrant ... was intended for use with Recorde's book.' Readers wishing to know more about the young king are referred to the *Oxford Dictionary of National Biography*.

MATHEMATICAL TERMINOLOGY

When setting out to introduce Euclidean geometry to English readers, Record found that the English language did not (at that time) contain many technical terms. Therefore he had two choices, he could either use longstanding Latin or Greek words and hope that they would become familiar, or he could invent new English words. He chose the latter course. So, for example, an equilateral triangle is

Introduction

a *threelike*, a square is a *likeside*, and a parallelogram is a *likejamme*. Similarly, acute, right and obtuse angles become in Recorde's terminology, *sharp, square and blunt* angles respectively. This list is not exhaustive and modern readers will have the pleasure of discovering Recorde's terminology for themselves in this volume. All his new words were the result of a thoughtful endeavour to anglicise mathematical language, with the laudable aim of minimising difficulties for those learning the subject through the medium of the English language for the first time. Unfortunately, Recorde's terms did not survive the passage of time and consequently to this day, schoolchildren have to wrestle with difficult Latin words like tangent instead of his much more homely and easily understood *touch line*.

SOURCES

Readers wanting to know more about Robert Recorde and his famous series of mathematical books should consult the following:

For an easily accessible biography visit the MacTutor History of Mathematics, 'Robert Recorde', [online] http://www-history.mcs.st-andrews.ac.uk/Biographies/Recorde.html. Written sources are:

Stephen Johnston, 'Recorde, Robert (c1512–1558)' *Oxford Dictionary of National Biography*, Oxford University Press, 2004.

Howell Lloyd, 'Famous in the Field of Number and Measure: Robert Record, Renaissance Mathematician', *Welsh History Review*, Vol. 2 (2000), pp. 254-282.

William Barr, 'A World View of Robert Recorde: A Brief Study of Tudor Cosmology, *Albion: A Quarterly Journal Concerned with British Studies*, Vol. 1, No. 1 (1969), pp. 1-9.

Joy B. Easton, 'The Early Editions of Robert Recorde's Ground of Artes', *Isis*, Vol. 58, No. 1 (Winter 1967), pp. 515-532.

Joy B. Easton, 'On the date of Robert Recorde's birth', *Isis*, Vol. 57, No. 1 (Spring 1966), p. 121.

Margaret E. Baron, 'A Note on Robert Recorde and the Dienes Blocks', *The Mathematical Gazette*, Vol. 50, No 374 (Dec 1966), pp. 363-369.

Louise Diehl Patterson, 'Recorde's Cosmography, 1556', *Isis*, Vol. 42, No. 3 (Oct 1951), pp. 208-218.

E.R. Sleight, 'Early English Arithmetics', *National Mathematics Magazine*, Vol. 16, No. 4 (Jan 1942), pp. 198-215 and Vol. 16 No. 5 (Feb 1942), pp. 243-251.

Introduction

Francis R. Johnson & Stanford V. Larkey, 'Robert Recorde's Mathematical Teaching and the Anti-Aristotelian Movement', *The Huntingdon Library Bulletin,* No. 7 (Apr 1935), pp. 59-87.

David Eugene Smith & Frances Marguerite Clarke, 'New Light on Robert Recorde', *Isis,* Vol. 8, No. 1 (Feb 1926), pp. 50-70.

David Eugene Smith, 'New Information Respecting Robert Recorde', *The American Mathematical Monthly,* Vol. 28, No. 8/9 (Aug–Sep 1921), pp. 296-300.

Frank V. Morley, 'Finis Coronat Opus', *The Scientific Monthly,* Vol. 10, No. 3 (Mar 1920), pp. 306-308.

HERE BEGINS
THE
PATHWAY TO KNOWLEDGE

The pathway to

KNOWLEDG, CONTAI-

NING THE FIRST PRIN=

ciples of Geometrie, as they
may moſte aptly be applied vn=
to pꝛactiſe, bothe for vſe of
inſtrumentes Geome=
tricall, and aſtrono=
micall and
alſo for pꝛoiection of plattes in euerye
kinde, and therfoꝛe much ne=
ceſſary for all ſortes of
men.

Geometries verdicte

All freſſhe fine wittes by me are filed,
All groſſe dull wittes wiſhe me exiled:
Thoughe no mannes witte reiect will I,
Yet as they be, I wyll them trye.

The argumentes of the foure bookes

The firſt booke declareth the definitions of the termes and names vſed in Geometry, with certaine of the chiefe grounds whereon the arte is founded. And then teacheth thoſe concluſions, which may ſerue diuerſely in al workes Geometricall.

The ſecond book doth ſette forth the Theoremes, (whiche maye be called approued truthes) ſeruinge for the due knowledge and ſure proofe of all concluſions and workes in Geometrye.

The third booke intreateth of diuers formes, and ſondry protractions thereto belonging, with the vſe of certain concluſions.

The fourth booke teacheth the right order of meaſuringe all platte formes, and bodies alſo, by reſon Geometricall.

TO THE GENTLE READER

XCVSE ME, GENTLE RE-
der if oughte be amiſſe, ſtra=
ung paths ar not trodē al tru
ly at the firſt: the way muſte
needes be comberous, wher
none hathe gone before. where no man hathe
geuen light, lighte is it to offend, but when the
light is ſhewed ones, light is it to amende. If
my light may ſo light ſome other, to eſpie and
marke my faultes, I wiſh it may ſo lighten thē,
that they may voide offence. Of ſtaggeringe
and ſtomblinge, and vnconſtaunt turmoilinge :
often offending, and ſeldome amending, ſuch
vices to eſchewe, and their fine wittes to ſhew
that they may winne the praiſe, and I to hold
the candle, whileſt they their glorious works
with eloquence ſette forth, ſo cunningly inuen
ted, ſo finely indited, that my bokes maie ſeme
worthie to occupie no roome. For neither is mi
wit ſo finelie filed, nother mi learning ſo larg-
ly lettred, nother yet mi laiſer ſo quiet and vn
cōbered, that I maie perform iuſtlie ſo learned
a laboure or accordinglie to accompliſhe ſo

r.ii.　　　　　　　　　　*haulte*

TO THE READER

haulte an enforcement, yet maie I thinke thus :
This candle did I light : this lighte haue I kin=
deled : that learned men maie ſe, to practiſe
their pennes, their eloquence to aduaunce, to
regiſter their names in the booke of memorie
I drew the platte rudelie, whereon thei maie
builde, whom god hath indued with learning
and liuelihod. For liuing by laboure doth learn-
ing ſo hinder, that learning ſerueth liuinge,
whiche is a peruers trade. Yet as carefull fa=
milie ſhall ceaſe hir cruell callinge, and ſuffre
anie laiſer to learninge to repaire, I will not
ceaſe from trauaile the pathe ſo to trade, that
finer wittes maie faſhion themſelues with ſuch
glimſinge dull light, a more complete woorke
at laiſer to finiſſhe, with inuencion agreable,
and aptnes of eloquence.

And this gentle reader I hartelie proteſt
where erroure hathe happened I wiſſhe it
redreſt.

TO THE MOST NO=
ble and puiſſaunt prince Edwarde the
ſirte by the grace of God, of En=
gland Fraunce and Ireland kynge, de=
fendour of the faithe, and of the
Churche of England and Ire=
lande in earth the ſu=
pzeme head.

T IS NOT VNKNO=
wen to youre maieſtie, moſte
ſoueraigne lozde, what great
diſceptacion hath been amon
geſt the wyttie men of all na=
cions, foz the exacte knowe=
ledge of true felicitie, bothe
what it is, and wherin it con
ſiſteth : touchynge whiche
thyng, their opinions al=
moſte were as many in num=
bze, as were the perſons of them, that either diſputed oz
wzote therof. But and if the diuerſitie of opinions in the
vulgar ſozt foz placyng of their felicitie ſhall be conſide=
red alſo, the varietie ſhall be found ſo great, and the opi=
nions ſo diſſonant, yea plainly monſterouſe, that no ho=
neſt witte would voucheſafe to loſe time in hearyng thē,
oz rather (as J may ſaie) no witte is of ſo exact remem=
bzance, that can conſider together the monſterouſe mul=
titude of them all. And yet not withſtādyng this repug=
nant diuerſitie, in two thynges do they all agree. Firſt
all do agre, that felicitie is and ought to be the ſtop and
end of all their doynges, ſo that he that hath a full de=
ſire to any thynge, how ſo euer it be eſtemed of other mē,
yet he eſtemeth himſelf happie, if he maie obtain it : and
contrary waies vnhappie if he can not attaine it. And
therfoze do all men put their whole ſtudie to gette that
thyng, wherin they haue perſwaded them ſelf that feli=
r.iii.   citie,

AN EPISTLE

citie doth confiſt. Wherfoze ſome whiche put their felici=
tie in fedyng their bellies, thinke no pain to be hard, noz
no dede to be vnhoneſt, that may be a meanes to fill that
foule panche. Other which put their felicitie in play and
ydle paſtimes, iudge no time euill ſpent, that is employ=
ed therabout : noz no fraude vnlawfull that may further
their winning. If J ſhould particularly ouerrune but the
common ſoztes of men, which put their felicitie in their
deſires, it wold make a great boke of itſelf. Therfoze wyl
J let them al go, and conclude as J began, That all men
employ their whole endeuour to that thing, wherin thei
thinke felicitie to ſtand. whiche thyng who ſo liſteth to
mark exactly, ſhall be able to eſpie and iudge the naturs
of al men, whoſe conuerſaciō he doth know, though thei
vſe great diſſimulacion to colour their deſires, eſpecially
whē they perceiue other men to miſlyke that which thei
ſo much deſire : Foz no mā wold gladly haue his appetite
impzroued. And herof cōmeth that ſeconde thing wherin
al agree, that euery man would moſt gladly win all other
men to his ſect, and to make thē of his opinion, and as
far as he dare, will diſpzaiſe all other mens iudgemētes,
and pzaiſe his own waies only, onles it be when he diſſi=
muleth, and that foz the furtherāce of his own purpoſe.
And this pzopertie alſo doth geue great light to the full
knowledge of mens natures, which as all men ought to
obſerue, ſo pzinces aboue other haue moſt cauſe to mark
foz ſundrie occaſions which may lye them on, wherof J
ſhall not nede to ſpeke any farther, conſideryng not only
the greatnes of wit, and exactes of iudgement whiche
god hath lent vnto your highnes perſon, but alſo ẏ moſt
graue wiſdom and pzofoūd knowledge of your maieſties
moſt honozable coūcel, by whō your highnes may ſo ſuf=
ficiently vnderſtād all thinges conuenient, that leſſe ſhal
it nede to vnderſtand by pziuate readyng, but yet not vt=
terly to refuſe to read as often as occaſion may ſerue, foz
bokes dare ſpeake, when men feare to diſpleaſe. But to
                                                    returne

TO THE KINGES MA.

returne agayne to my firſte matter, if none other good
thing maie be lerned at their maners, which ſo wꝛ̃ogful=
ly place their felicity, in ſo miſerable a cõditiõ (that while
they thinke them ſelfes happy, their felicitie nuſt nedes
ſeme vnluckie, to be by them ſo euill placed) yet this may
men learn at them, by thoſe .ii. ſpectacles to eſpye the ſe=
crete natures and diſpoſitions of others, whiche thyng
vnto a wiſe man is muche auailable. And thus will J o=
mit this great tablement of vnhappie hap, and wil come
to .iii. other ſoꝛtes of a better degre, wherof the one put=
teth felicitie to conſiſt in power and royaltie. The ſecond
ſoꝛte vnto power annexeth woꝛldly wiſdome, thinkyng
him full happie, that could attayn thoſe two, wherby he
might not onely haue knowledge in all thynges, but al=
ſo power to bꝛyng his deſires to ende. The thyꝛd ſoꝛt e=
ſtemeth true felicitie to conſiſt in wyſdom annexed with
vertuouſe maners, thinkyng that they can take harme
of nothyng, if they can with their wyſedome ouercome
all vyces. Of the firſte of thoſe three ſoꝛtes there hath
been a great numbꝛe in all ages, yea many mightie kin=
ges and great gouernoures, whiche cared not greately
howe they myght atchieue their pourpoſe, ſo that they
dyd pꝛeuayle : noꝛ did not take any greatter care foꝛ go=
uernance, then to kepe the people in onely feare of them.
Whoſe common ſentence was alwaies this : Oderint
dum metuant. And what good ſucceſſe ſuche menne
had, all hiſtoꝛies doe repoꝛt. Yet haue they not wanted
excuſes : yea Julius Caeſar (whiche in dede was of the
ſecond ſoꝛte) maketh a kynde of excuſe by his common
ſentence, foꝛ them of that fyꝛſte ſoꝛte, foꝛ he was euer
woonte to ſaie : εἴπερ γὰρ ἀδικείν χη , τυραννίδΘ-
πέρι κάλλισον ἀδικείν , τ᾽ ἄλλα δ᾽ ευσεβείν χεών .
Whiche ſentence J wyſſhe had neuer been learned out of
Gꝛecia. But now to ſpeake of the ſecond ſoꝛt, of whiche
there hathe been verye many alſo, yet foꝛ this pꝛeſent
time amongeſt them all, J wyll take the exaumples of
                                                    kyng

AN EPISTLE

kynge Phylippe of Macedonie, and of Alexander his
sonne, that valiaunt conquerour. First of kinge Phylip
it appeareth by his letter sente vnto Aristotle that fa=
mous philosopher, that he more delited in the birthe of
his sonne, for the hope of learning and good education,
that might happen to him by the said Aristotle, then he
didde reioyse in the continuaunce of his succession, for
these were his wordes and his whole epistle, worthye to
bee remembred and registred euery where.

Φιλιππ℗ Αρισοΐελΐ χαίρειν.

ἴδι μοι γεγονότα ὑόμ. πολλὰμ οὐμ τοῖσ θεοῖσ χάριμ ἔχω,
ὀυχ ὅυτωσ ἐπί τῆ γεννήσει τ παιδόσ, ὡσ ἐπί τῷ κα=
τὰ τὴμ σὴμ ἡλικίαμ αὐτὸμ γεγονέναι ἐλπίζω γὰρ
αὐτὸμ ὑπὸ σὖ ῥαφέντα ὶϳ παιδευθέντα ἄξιομ ἴσεσθαι ὶϳ
ἡμῶμ ὶϳ τῆσ τῶμ πραγμάτωμ διαδοχῆσ.

That is thus in sense,

Philip vnto Aristotle sendeth gretyng.

You shall vnderstande, that I haue a sonne borne, for
whiche cause I yelde vnto God moste hartie thankes,
not so muche for the byrthe of the childe, as that it was
his chaunce to be borne in your tyme. For my trust is,
that he shall be so brought vp and instructed by you, that
he shall become worthie not only to be named our sonne,
but also to be the successour of our affayres.

And his good desire was not all vayne, for it appered
that Alexander was neuer so busied with warres (yet
was he neuer out of moste terrible battaile) but that in
the middes therof he had in remembraunce his studies,
and caused in all countreies as he went, all strange bea=
                                                         stes,

TO THE KINGES MA.

ſtes, foꝛules and fiſhes, to be taken and kept foꝛ the ayd
of that knouledg, which he learned of Ariſtotle : And al=
ſo he had with him alwayes a greate numbꝛe of learned
men. And in the moſte buſye tyme of all his warres a=
againſt Darius kinge of Perſia, when he harde that A=
riſtotle had putte foꝛthe certaine bookes of ſuche know=
ledge wherein he hadde befoꝛe ſtudied, hee was offended
with Ariſtotle, and wꝛote to hym this letter.

Αλέζανδꝛ⊙ Αρισοτέλει εὖ πράτ̄ειμ.

Ὀυκ ὀρθῶσ ἐπόιησασ ἐκδ᾽ους τοὺσ ἀκροαμαᾶικούσ τῶμ
λόγωμ, τίνι γὰρ διοισομὄμ ἡμεῖσ τῶμ ἄλλωμ, ἐι καθ᾽ ους
ἐπαιδεύθημδμ λόγουσ, ὄυτοι πάντωμ ἐσουῖαι κοινόι, ἐγὼ
δὲ βૅλόι μὴμ ἄμ τᾶισ περι τὰ ἀριστα ἐμπειρίαισ, ἢ τᾶισ
δ᾽ωάμεσι διαφέριμ. ἔρρωσο.   that is

Alexander vnto Ariſtotle ſendeth greeting.

You haue not doone well, to put foꝛthe thoſe bﬦkes
of ſecrete phyloſophy intituled, ἀκροαμαᾶικόι. Foꝛ wher=
in ſhall we excell other, yf that knowledge that wee haue
ſtudied, ſhall be made commen to all other men, namely
ſithe our deſire is to excelle other men in experience and
knowledge, rather then in power and ſtrength. Farewell.

By whyche lettre it appeareth that hee eſtemed lear=
ninge and knowledge aboue power of men. And the like
iudgement did he vtter, when he beheld the ſtate of Di=
ogenes Linicus, adiudginge it the beſte ſtate next to his
owne, ſo that he ſaid : If I were not Alexander, I wolde
wiſhe to be Diogenes. Whereby apeareth, how he eſtee=
med learning, and what felicity he putte therin, reputing
al the woꝛlde ſaue him ſelfe to be inferiour to Diogenes.
And bi al coniecturs, Alexander did eſteme Diogenes one
of them whiche contemned the vaine eſtimation of the
diſceitfull woꝛld, and put his whole felicity in knowledg
of vertue, and pꝛactiſe of the ſame, though ſome repoꝛte

AN EPISTLE

that he knew moze vertue then he folowed : But whatſo
euer he was, it appeareth that Socrates and Plato and
many other did forſake their liuings and ſel away their
patrimony, to the intent to ſeek and trauaile foz lear=
ning, which examples I ſhall not need to repete to pour
Maieſty, partly foz that pour highnes doth often reade
them and other lyke, and partly ſith pour maieſty hath
at hand ſuch learned ſchoolemapſters, which can much
better thē I, declare them vnto pour highnes, and that
moze largely alſo then the ſhoztenes of thys epiſtle will
permit. But thys may I pet adde, that King Solomon
whoſe renoume ſpzed ſo farre abzoad, was very greatlye
eſtemed foz his wonderfull power and exceading trea=
ſure, but pet much moze was he eſtemed foz his wiſdom
And him ſelfe doth bear witnes, that wiſedom is better
then pzecious ſtones. pea all thinges that can be deſi=
red ar not to be compared to it. But what needeth to al=
ledge one ſentence of him, whoſe bookes altogither do
none other thing, then ſet fozth the pzaiſe of wiſedom &
knowledg? And his father king Dauid iopneth uertuous
conuerſacion and knowledg togither, as the ſumme of
perfection and chief felicity. Wherfoze I maye iuſtelpe
conclude, that true felicity doth conſiſt in wiſdome and
vertu. Then if wiſdome be as Cicero defineth it, Diui=
narum atq; humanarum rerum ſcientia, then ought
all men to trauail foz knowledg in matters both of reli=
gion and humaine doctrine, if he ſhall be counted wpſe,
and able to attaine true felicitie : But as the ſtudy of re=
ligious matters is moſt pzincipall, ſo I leue it foz this
time to them that better can wzite of it then I can. And
foz humaine knowledge thys wil I boldlp ſay, that who
ſoeuer wpll attain true iudgment therin, muſt notenlp
trauail in y̆ knowledg of the tungs, but muſt alſo befoze
al other arts, taſte of the mathematical ſciences, ſpecial
lp Arithmetike and Geometry, without which it is not
poſſible to attapn full knowledg in any art. Which may
ſufficietly by gathered by Ariſtotle not ōlp in his bookes
of de=

TO THE KINGES MA.

of demonstration (whiche can not be vnderstand with=
out Geometry) but also in all his other workes. And be=
fore him Plato his master wrote this sentence on his
schole house dore.   Α γεωμέτρητος ονδεισ εισίτω.   Let
no man entre here (saith he) without knowledg in Ge=
ometry. Wherfore moste mighty prince, as your most
excellent Maiesty appeareth to be borne vnto most per=
fect felicity, not only by reaso that God moued with the
long prayers of this realme, did send your highnes as
a moste comfortable inheritour to the same, but also in
that your Maiesty was borne in the time of such skilful
schoolmaisters & learned techers, as your highnes doth
not a little reioyse in, and profite by them in all kind of
vertu & knowledg. Amogst which is that heauely know
ledg most worthely to be praised, wherbi the blindnes of
errour & superstition is exiled, & good hope coceiued that
al the sedes & fruts therof, with all kindes of vice & ini=
quite, wherby vertu is hindered, & iustice defaced, shal be
clean extrirped and rooted out of this realm, which hope
shal increase more and more, if it may appear that lear=
ning be estemed & florish within this realm. And al be it
the chief learnig be the diuine scriptures, which instruct
the mind principally, & nexte therto the lawes politike,
which most specially defed the right of goodes, yet is it
not possible that those two can long be wel vsed, if that
ayde want that gouerneth health and expelleth sicknes,
which thing is done by Physik, & these require the help
of the .vii. liberall sciences, but of none more then of A=
rithmetik and Geometry, by which not only great thin
ges ar wrought touchig accoptes in al kinds, & in siruai=
yng & measuring of lades, but also al arts depend partly
of the, & building which is most necessary can not be w=
out them, which thing cosidering, moued me to help to
serue your maiesty in this point as wel as other wais, &
to do what mai be in me, y not oly thei which studi prici
palli for lernig, mai haue furderace bi mi poore help, but

AN EPISTLE

alſo thoſe whiche haue no tyme to trauaile foʒ exacter
knowledge, may haue ſome helpe to vnderſtand in thoſe
Mathematicall artes, in whiche as J haue all readye
ſet foʒth ſumwhat of Arithmetike, ſo god willing J in
tend ſhoʒtly to ſet foʒth a moʒe exacter woʒke therof. And
in the meane ceaſon foʒ a taſte of Geometry, J haue
ſette foʒthe this ſmall introduction, deſiring your grace
not ſo muche to beholde the ſimplenes of the wooʒke, in
compariſon to your Maieſties excellencye, as to fa=
uour the edition therof, foʒ the ayde of your humble
ſubiectes, which ſhal thinke them ſelues moʒe and moʒe
dayly bounden to your highnes, if when they ſhall per=
ceaue your graces deſyʒe to haue theym pʒofited in all
knowledge and vertue. And J foʒ my pooʒe ability con
ſidering your Maieſties ſtudye foʒ the increaſe of lear=
ning generally through all your highenes dominions,
and namely in the vniuerſities of Oxfoʒde and Came=
bʒidge, as J haue an earneſt good will as far as my ſim=
ple ſeruice and ſmall knowledg will ſuffice, to helpe to=
ward the ſatiſfiyng of your graces deſire, ſo if J ſhall
perceaue that my ſeruice may be to your maieſties con=
tētacion, J wil not only put foʒth the other two books,
whiche ſhoulde haue beene ſette foʒth with theſe two, yf
miſfoʒtune had not hindered it, but alſo J wil ſet foʒth
other bookes of moʒe exacter arte, bothe in the Latine
tongue and alſo in the Englyſhe, whereof parte bee all
readye wʒitten, and newe inſtrumentes to theym deui=
ſed, and the reſidue ſhall bee eanded with all poſſible
ſpeede. J was boldened to dedicate this booke of Ge=
ometrye vnto your Maieſtye, not ſo muche bycauſe it
is the firſte that euer was ſette foʒthe in Engliſhe, and
therefoʒe foʒ the noueltye a ſtraunge pʒeſente, but foʒ
that J was perſwaded, that ſuche a wyſe pʒince doothe
deſire to haue a wiſe ſoʒte of ſubiectes. Foʒ it is a kyn=
ges chiefe reioyſinge and gloʒy, if his ſubiectes be riche
in ſubſtaunce, and wytty in knowledge : and contrarye
waies

TO THE KINGES MA.

waies nothyng can bee moze greuouſe to a noble kyng,
then that his realme ſhould be other beggerly oz full of
ignozaunce : But as god hath geuen your grace a realme
bothe riche in commodities and alſo full of wyttie men,
ſo J truſte by the readyng of wyttie artes (whiche be as
the whette ſtones of witte) they muſte needes increaſe
moze and moze in wyſedome, and peraduenture kynde
ſome thynge towarde the ayde of their ſubſtaunce,
whereby your grace ſhall haue newe occaſion to reioyce,
ſeyng your ſubiectes to increaſe in ſubſtance oz wiſdom,
oz in both. And thei again ſhal haue new and new cauſes
to pzay foz your maieſtie, perceiuyng ſo gracĳouſe a mind
towarde their benefite. And J truſte (as J deſire) that a
great numbze of gentlemen, eſpecially about the courte,
whiche vnderſtand not the latin tong, oz els foz the hard
neſſe of the mater could not away with other mens wzi=
tyng, will fall in trade with this eaſie fozme of teachyng
in their vulgar tong, and ſo employe ſome of their tyme
in honeſt ſtudie, whiche were wont to beſtowe moſt part of
their time in triflyng paſtime : Foz vndoubtedly if the
mean other your maieſties ſeruice, other their own wiſ=
dome, they will be content to employ ſome tyme aboute
this honeſt and wittie exerciſe. Foz whoſe encouragemēt
to the intent they maie perceiue what ſhall be the vſe of
this ſcience, J haue not onely wzitten ſomewhat of the
vſe of Geometrie, but alſo J haue annexed to this boke
the names and bzefe argumentes of thoſe other bokes
whiche J will ſet fozth hereafter, and that as ſhoztly as
it ſhall appeare vnto your maieſtie by coniecture of their
diligent vſyng of this firſt boke, that they wyll vſe well
the other bokes alſo. In the meane ceaſon, and at all ti=
mes J wil be a continuall petitioner, that god may wozk
in all engliſhe hartes an erneſt mynde to all honeſt exer=
ciſes, wherby thei may ſerue the better your maieſtie and

s.iii.   the

TO THE KINGES MA.

the realm. And for your highnes J befech the moſt mer=
cifull god, as he hath moſt fauourably ſent you vnto vs,
as our chefe comforter in earthe, ſo that he will increaſe
your maieſtie daiely in all vertue and honor with moſte
proſperouſe ſucceſſe, and augment in vs your moſt hum=
ble ſubiectes, true loue to godward, and iuſt obedience to=
ward your highnes with all reuerence and ſubiection.

At London the .xxviij. daie of Januarie.   M.D.L.I.

Your maieſties moſte humble ſer=
uant and obedient ſubieⅽt,
Robert Recorde.

THE PREFACE

Declaring briefely the commodi= ties of Geometrye, and the necessitye thereof.

Eometrye may thinke it selfe to
suſtaine great iniury, if it ſhalt be
inforced other to ſhow her mani
fold commodities, or els not to
preaſe into the ſight of men, and
therefore might this wayes an=
ſwere briefely : Other I am able
to do you much good, or els but
litle. If I bee able to doo you
much good, then be you not your owne friendes, but greatlye
your owne enemies to make ſo little of me, which maye pro=
fite you ſo muche. For if I were as vncurteous as you vnkind,
I ſhuld vtterly refuſe to do them any good, which will ſo cu=
riouſly put me to the trial and profe of my commodities, or els
to ſuffre exile, and namely ſithe I ſhal only yeld benefites to
other, and receaue none againe. But and if you could ſaye
truely, that my benefites be nother many nor yet greate, yet if
they bee anye, I doo yelde more to you, then I doo receaue
againe of you, and therefore I oughte not to bee repelled of
them that loue them ſelfe, althoughe they loue me not at all
for my ſelfe. But as I am in nature a liberall ſcience, ſo canne
I not againſte nature contende with your inhumanitye, but
muſte ſhewe my ſelfe liberall euen to myne enemies. Yet this
is my comforte againe, that I haue none enemies but them that
knowe me not, and therefore may hurte themſelues, but can
not noye me. If they diſpraiſe the thinge that they know not,
all wiſe men will blame them and not credite them. and yf
they thinke they knowe me, lette theym ſhewe one vntruthe
and erroure in me, and I wyll geue the victorye.

Yet

THE PREFACE

*Yet can no humayne science saie thus, but I onely, that there is
no sparke of vntruthe in me : but all my doctrine and workes
are without any blemishe of errour that mans reason can dif=
cerne. And nexte vnto me in certaintie are my three systers,
Arithmetike, Musike, and Astronomie, whiche are also so
nere knitte in amitee, that he that loueth the one, can not de=
spise the other, and in especiall Geometrie, of whiche not on=
ly these thre, but all other artes do borow great ayde, as part=
ly hereafter shall be shewed. But first will I beginne with the
vnlearned sorte, that you maie perceiue how that no arte can
stand without me. For if I should declare how many wayes
my helpe is vsed, in measuryng of ground, for medow, corne,
and wodde : in hedgyng, in dichyng, and in stackes makyng, I
thinke the poore Husband man would be more thankefull vn=
to me, then he is nowe, whyles he thinketh that he hath small
benefite by me. Yet this maie be coniecture certainly, that if
he kepe not the rules of Geometrie, he can not measure any
ground truely. And in dichyng, if he kepe not a proportion of
bredth in the mouthe, to the bredthe of the bottome, and iuste
slopenesse in the sides agreable to them bothe, the diche shall be
faultie many waies. When he doth make stackes for corne,
or for heye, he practiseth good Geometrie, els would thei not
long stand : So that in some stackes, whiche stand on foure pil=
lers, and yet made round, doe increase greatter and greatter a
good height, and then againe turne smaller and smaller vnto
the toppe : you maie see so good Geometrie, that it were very
difficult to counterfaite the lyke in any kynde of buildyng. As
for other infinite waies that he vseth my benefite, I ouerpasse
for shortnesse.*

*Carpenters, Karuers, Joyners, and Masons, doe willingly
acknowledge that they can worke nothyng without reason of
Geometrie, in so muche that they chalenge me as a peculiare
science for them. But in that they should do wrong to all other
men, seyng euerie kynde of men haue som benefit by me, not on
ly in buildyng, whiche is but other mennes costes, and the arte
of Carpenters, Masons, and the other aforesayd, but in their*

own

THE PREFACE

owne priuate profeſſion, wherof to auoide tedioufnes I
make this reherfall.
Sith Merchauntes by fhippes great riches do winne,
    I may with good righte at their feate beginne.
The Shippes on the fea with Saile and with Ore,
    were firfte founde, and ftyll made, by Geometries lore,
Their Compas, their Carde, their Pulleis, their Ankers,
    were founde by the ſkill of witty Geometers.
To fette forth the Capftocke, and eche other parte,
    wold make a greate fhowe of Geometries arte.
Carpenters, Caruers, Ioiners and Mafons,
    Painters and Limners with fuche occupations,
Broderers, Goldefmithes, if they be cunning,
    Muft yelde to Geometrye thankes for their learning.
The Carte and the Plowe, who doth them well marke,
    Are made by good Geometrye. And fo in the warke
Of Tailers and Shoomakers, in all fhapes and fafhion,
    The woorke is not praifed, if it wante proportion.
So weauers by Geometrye hade their foundacion,
    Their Loome is a frame of ftraunge imaginacion.
The wheele that doth fpinne, the ftone that doth grind,
    The Myll that is driuen by water or winde,
Are workes of Geometrye ftraunge in their trade,
    Fewe could them deuife, if they were vnmade.
And all that is wrought by waight or by meafure,
    without proofe of Geometry can neuer be fure.
Clockes that be made the times to deuide,
    The wittieft inuencion that euer was fpied,
Nowe that they are common they are not regarded,
    The artes man contemned, the woorke vnrewarded.
But if they wre fcarfe, and one for a fhewe,
    Made by Geometrye, then fhoulde men know,
That neuer was arte fo wonderfull witty,
    So needefull to man, as is good Geometry.
The firfte findinge out of euery good arte,
    Seemed then vnto men fo godly a parte,

                                    t.i.   *That*

THE PREFACE

That no recompence might fatiffy the finder,
  But to make him a god, and honoure him for euer.
So Ceres and Pallas, and Mercury alfo,
  Eolus and Neptune, and many other mo,
Were honoured as goddes, bicaufe they did teache,
  Firfte tillage and weuinge and eloquent fpeache,
Or windes to obferue, the feas to faile ouer,
  They were called goddes for their good indeuour.
Then were men more thankefull in that golden age:
  This yron worlde nowe vngratefull in rage,
Wyll yelde the thy reward for trauaile and paine,
  with fclaunderous reproch, and fpitefull difdaine.
Yet thoughe other men vnthankfull will be,
  Suruayers haue caufe to make muche of me.
And fo haue all Lordes, that landes do poffeffe:
  But Tennauntes I feare will like me the leffe.
Yet do I not wrong but meafure all truely,
  And yelde the full right to euerye man iuftely.
Proportion Geometricall hath no man oppreft,
  Yf anye bee wronged, I wifhe it redreft.

      But now to procede with learned profeffions, in Lo=
gike and Rhetorike and all partes of phylofophy, there nea=
deth none other proofe then Ariftotle his teftimony, whiche
without Geometry proueth almoft nothinge. In Logike all
his good fyllogifmes and demonftrations, hee declareth by the
principles of Geometrye. In philofophye, nether motion,
nor time, nor ayrye impreffions coulde hee aptely declare,
but by the helpe of Geometrye as his woorkes do witnes. Yea
the faculties of the minde dothe hee expreffe by fimilitude to
figures of Geometrye. And in morall phylofophy he thought
that iuftice coulde not wel be taught, nor yet well executed
without proportion geometricall. And this eftimacion of Ge=
ometry he maye feeme to haue learned of his maifter Plato,
which withoute Geometrye wolde teache nothinge, nother
wold admitte any to heare him, except he were experte in
Geometry. And what meruaile if he fo muche eftemed geome=
trye, feinge his opinion was, that Godde was alwaaies wor=
                                                              kinge

THE PREFACE

kinge by Geometrie? Whiche fentence Plutarche declareth at
large. And although Plato do vfe the helpe of Geometrye in
all the moft waighte matter of a common wealth, yet it is fo
generall in vfe, that no fmall thinges almoft can be wel done
without it. And therfore faith he : that Geometrye is to be lear
ned, if it were for none other caufe, but that all other artes are
bothe foner and more furely vnderftand by helpe of it.

What greate help it dothe in phyfike, Galene doth fo often
and fo copioufely declare, that no man whiche hath redde any
booke almofte of his, can be ignorant thereof in fo much that
he coulde neuer cure well a round vlcere, tyll reafon geo=
metricall dydde teache it hym. Hippocrates is earneft in ad=
monyfhynge that ftudy of geometrie muft prepare the way to
phyfike, as well as to all other artes.

I fhoulde feeme fomewhat to tedious, if I fhoulde recken
vp, howe the diuines alfo in all their myfteries of fcripture
doo vfe healpe of geometrie : and alfo that lawyers can ne=
uer vnderftande the hole lawe, no nor yet the firfte title ther
of exactly without Geometrie. For if lawes can not well be
eftablifhed, nor iuftice duelie executed without geometricall
proportion, as bothe Plato in his Politike bokes, and Ariftotle
in his Moralles doo largely declare. Yea fithe Lycurgus that
cheefe lawmaker amongeft the Lacedemonians, is mofte
praifed for that he didde chaunge the ftate of their common
wealthe frome the proportion Arithmeticall to a proportion
geometricall, whiche without knowledg of bothe he coulde
not dooe, than is it eafye to perceaue howe neceffarie Geo=
metrie is for the lawe and ftudentes therof. And if I fhall
faie precifelie and freelie as I thinke, he is vtterlie deftitute
of all abilitee to iudge in anie arte, that is not fomme what
experte in the Theoremes of Geometrie.

And that caufed Galene to fay of hymfelfe, that he coulde ne=
uer perceaue what a demonftration was, no not fo muche, as
whether there were any or none, tyll he had by geometrie
gotten abilitee to vnderftande it, although he heard the befte
teachers that were in his tyme. It fhuld be to longe and nede=

t.ii.

leffe

THE PREFACE

*leſſe alſo to declare what helpe all other artes Mathemati=
call haue by geometrie, ſith it is the grounde of all theyr cer=
teintie, and no man ſtudious in them is ſo doubtful therof, that
he ſhall nede any perſuaſion to procure credite thereto. For he
can not reade .ij. lines almoſte in any mathematicall ſcience,
but he ſhall eſpie the nedefulnes of geometrie. But to auoyde
tedtouſneſſe I will make an ende hereof with that famous
ſentence of auncient Pythagorus, That who ſo will trauayle
by learnyng to attayne wyſedome, ſhall neuer approche to
any excellencie without the artes mathematicall, and eſpeci=
ally Arithmetike and Geometrie.*

*And yf I ſhall ſomewhat ſpeake of noble men, and gouer=
nours of realmes, howe needefull Geometrye maye bee vn=
to them, then muſt I repete all that I haue ſayde before, ſithe
in them ought all knowledge to abounde, namely that maye
appertaine either to good gouernaunce in time of peace, ey=
ther wittye pollicies in time of warre. For miniſtration of
good lawes in time of peace Lycurgus example with the teſti
monies of Plato and Ariſtotle may ſuffice. And as for war=
res, I might thinke it ſufficient that Vegitius hath written,
and after him Valturius in commendation of Geometry, for
vſe of warres, but all their woordes ſeeme to ſaye nothinge,
in compariſon to the example of Archimedes worthy woor=
kes made by geometrie, for the defence of his countrey, to
reade the wonderfull praiſe of his wittie deuiſes, ſet foorthe
by the moſte famous hyſtories of Liuius, Plutarche, and
Plinie, and all other hyſtoriographiers, whyche wryte
of the ſtronge ſiege of Syracuſæ made by that vali=
ant capitayne, and noble warriour Marcellus, whoſe po=
wer was ſo great, that all men meruayled how that one ci=
tee coulde withſtande his wonderfull force ſo longe. But much
more woulde they meruaile, if they vnderſtode that one man
onely dyd withſtand all Marcellus ſtrength, and with coun=
ter engines deſtroied his engines to the vtter aſtonyſhment of
Marcellus, and all that were with hym. He had inuented
ſuche balaſtelas that dyd ſhoote out a hundred dartes at one*

shoote

THE PREFACE

ſhotte, to the great deſtruction of Marcellus *ſouldiours, wherby a fonde tale was ſpredde abrode, how that in Syracu= ſe there was a wonderfull gyant, whiche had a hundred han- des, and coulde ſhoote a hundred dartes at ones. And as this fable was ſpredde of Archimedes, ſo many other haue been fayned to bee gyantes and monſters, bycauſe they dyd ſuche thynges, which farre paſſed the witte of the common peo= ple. So dyd they feyne Argus to haue a hundred eies, bicauſe they herde of his wonderfull circumſpection, and thoughte that as it was aboue their capacitee, ſo it could not be, onleſſe he had a hundred eies. So imagined they Janus to haue two faces, one lokyng forwarde, and an other backwarde, bycauſe he coulde ſo wittily compare thynges paſte with thynges that were to come, and ſo duely pondre them, as yf they were all preſent. Of like reaſõ did they feyn Lynceus to haue ſuch ſharp ſyght, that he coulde ſee through walles and hylles, bycauſe peraduenture he dyd by naturall iudgement declare what cõ= moditees myght be digged out of the grounde. And an infi= nite noumbre lyke fables are there, whiche ſprange all of lyke reaſon.*

For what other thyng meaneth the fable of the great gyant Atlas, whiche was ymagined to beare vp heauen on his ſhul= ders? but that he was a man of ſo high a witte, that it reached vnto the ſkye, and was ſo ſkylfull in Aſtronomie, and coulde tell before hande of Eclipſes, and other like thynges as truely as though he dyd rule the ſterres, and gouerne the planettes.

So was Eolus accompted god of the wyndes, and to haue them all in a caue at his pleaſure, by reaſon that he was a wit tie man in naturall knowlege, and obſerued well the change of wethers, and was the fyrſt that taught the obſeruation of the wyndes. And lyke reſon is to be geuen of al the old fables.

But to retourne agayne to Archimedes, he dyd alſo by arte perſpectiue (whiche is a parte of geometrie) deuiſe ſuch glaſ= ſes within the towne of Syracuſe, that dyd bourne their en= emies ſhyppes a great way from the towne, whyche was a meruaylous politike thynge. And if I ſhulde repete the va=

riete

THE PREFACE

rietees of ſuche ſtraunge inuentions, as Archimedes and o=
thers haue wrought by geometrie, I ſhould not onely excede
the order of a preface, but I ſhould alſo ſpeake of ſuche thyn
ges as can not well be vnderſtande in talke, without ſomme
knowledge in the principles of geometrie.

But this will I promiſe, that if I may perceaue my pay=
nes to be thankfully taken, I wyll not onely write of ſuche
pleaſant inuentions, declaryng what they were, but alſo wil
teache howe a great numbre of them were wroughte, that
they may be practiſed in this tyme alſo. Wherby ſhallbe plain
ly perceaued, that many thynges ſeme impoſſible to be done,
whiche by arte may very well be wrought. And whan they
be wrought, and the reaſon therof not vnderſtande, than ſay
the vulgare people, that thoſe thynges are done by negroman=
cy. And hereof came it that fryer Bakon was accompted ſo
greate a negromancier, whiche neuer vſed that arte (by any
coniecture that I can fynde) but was in geometrie and other
mathematicall ſciences ſo experte, that he coulde dooe by
theim ſuche thynges as were wonderfull in the ſyght of moſt
people.

Great talke there is of a glaſſe that he made in Oxforde, in
whiche men myght ſee thynges that were doon in other pla=
ces, and that was iudged to be done by power of euyll ſpi=
rites. But I knowe the reaſon of it to bee good and naturall,
and to be wrought by geometrie (ſythe perſpectiue is a parte
of it) and to ſtande as well with reaſon as to ſee your face in
cōmon glaſſe. But this concluſion and other dyuers of lyke
ſorte, are more mete for princes, for ſundry cauſes, than for
other men, and ought not to bee taught commonly. Yet to
repete it, I thought good for this cauſe, that the worthynes of
geometry myght the better be knowen, & partly vnderſtanding
geuen, what wonderfull thynges may be wrought by it, and ſo
conſequently how pleaſant it is, and how neceſſary alſo.

And thus for this tyme I make an end. The reaſon of ſom
thynges done in this boke, or omitted in the ſame, you ſhall
fynde in the preface before the Theoremes.

The definitions of the principles of
GEOMETRY.

EOMETRY TEA=
*cheth the drawyng, Meafuring
and proporcion of figures, but
in as muche as no figure can bee
drawen, but it mufte haue cer=
tayne boūdes and inclofures of
lines : and euery lyne alfo is be=
gon and ended at fome certaine
prycke, fyrft it fhal be meete to
know thefe fmaller partes of e=*
*uery figure, that therby the whole figures may the better bee
iudged, and diftinČte in fonder.*

A Poynt or a Prycke, *is named of* Geometricians *that   A pointe.
fmall and vnfenfible fhape, whiche hath in it no partes, that
is to fay : nother length, breadth nor depth. But as this exaČt=
nes of definition is more meeter for onlye Theorike fpecula=
cion, then for praČtife and outwarde worke (confideringe
that myne intente is to applye all thefe whole principles to
woorke) I thynke meeter for this purpofe, to call a* poynt
or prycke, *that fmall printe of penne, pencyle, or other
inftrumente, whiche is not moued, nor drawen from his fyrft
touche, and therfore hath no notable length nor bredthe : as
this example doeth declare.*   ∴

*Where I haue fet .iiij. prickes, eche of them hauyng both
lēgth and bredth, thogh it be but fmal, and therfore not notable.*

Nowe of a great numbre of thefe prickes, is made a Lyne,
as you may perceiue by this forme enfuyng. ....
*where as I haue fet a numbre of prickes, fo if you with your
pen will fet in more other prickes betweene euerye two of
thefe, then wil it be a lyne, as here you may fee* ——— *and*
this lyne, *is called of* Geometricians, Lengthe withoute   A lyne.
breadth.
But as they in theyr theorikes (which ar only mind workes)
                              A   do

DEFINITIONS

*do precifely vnderftand thefe definitions, fo it fhal be fuffici=
ent for thofe men, whiche feke the vfe of the fame thinges,
as fenfe may duely iudge them, and applye to handy workes
if they vnderftand them fo to be true, that outwarde fenfe
canne fynde none erroure therin.*

    *Of lynes there bee two principall kyndes. The one is cal
led a right or ftraight lyne, and the other a croked lyne.*

    A Straight lyne, *is the fhorteft that maye be drawenne
betweene two prickes.*

    *And all other lines, that go not right forth from prick to
prick, but boweth any waye fuch are called* Croked lynes
*as in thefe examples folowyng ye may fe, where I haue fet
but one forme of a ftraight lyne, for more formes there be
not, but of crooked lynes there be innumerable diuerfities,
whereof for examples fum I haue fette here.*

———————————— *A right lyne.*

Croked   lynes

croked lines

    so now you muft vnderftand, that
euery lyne is drawen betwene
twoo prickes, *wherof the one is at*
the beginning, *and the other at the ende.*

    *Therfore when foeuer you do fee any
formes of lynes to touche at one notable
pricke, as in this example, then fhall you*

B

A C

not

GEOMETRICALL.

not call it one croked lyne, but rather twoo lynes : in as
muche as there is a notable and sensible angle by .A. whiche   An angle.
euermore is made by the meetyng of two seuerall lynes. And
likewayes shall you iudge of this figure,
whiche is made of two lines, and not of
one onely.

    So that whan so euer any suche meetyng of lines doth hap=
pen, the place of their metyng is called an Angle or corner.

    Of angles there by three generall kindes : a sharpe angle, a
square angle, and a blunte angle. The square angle, *whiche*   A righte
is commonly named a right corner, *is made of twoo lynes* angle.
meetyng together in fourme of a squire, whiche two lines, if
they be drawen forth in length, will crosse one an other : as in
the examples folowyng you maie see.

    A sharpe angle *is so called, becaufe it is leffer than is a*   A sharpe
square angle, and the lines that make it, do not open so wide in corner.
their departynge as in a square corner, and if thei be drawen
croffe, all fower corners will not be equall.

    A blunt or brode corner, *is greater then is a square an=*   A blunte
gle, and his lines do parte more in fonder then in a right angle, angle.
of whiche all take thefe examples.

Right angles.

And thefe angles (as
you fee) are made part=
ly of ftreght lynes, part=
ly of croked lines, and
partly of both together.
Howbeit in right angles

Sharpe angles.

I haue put none example of croked lines, becaufe it would
A.ij.   muche

DEFINITIONS

muche trouble a lerner | Blunte or brode angles
to iudge them : for their
true iudgment doth ap=
pertaine to arte perspe=
ctiue, and as I may fay,
rather to reafon then to fenfe.

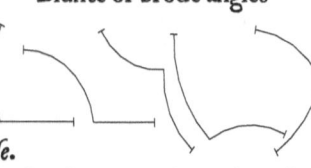

But now as of many prickes there is made one line, fo of diuerfe lines are there made fundry formes, figures, *and fhapes, whiche all yet be called by one propre name,* **A platte** Platte formes, *and thei haue bothe* length and bredth, **forme.** but yet no depeneffe.

And the boundes *of euerie platte forme are lines : as by the examples you maie perceiue.*

Of platte formes fome be plain, and fome be croked, and fome partly plaine, and partlie croked.

A plaine *A* plaine platte *is that, whiche is made al equall in height,* **platte.** *fo that the middle partes nother bulke vp, nother fhrink down more then the bothe endes.*

For whan the one parte is higher then the other, then is it na= **A crooked** med *a* Croked platte. **platte.**

And if it be partlie plaine, and partlie crooked, then is it called a Myxte platte, *of all whiche, thefe are exaumples.*

A plaine platte.   A croked platte.   *And as of many prickes is made a line, and of di= uerfe lines one platte forme, fo of manie plattes* **A bodie.** *is made a bo=* A myxte platte. die, *whiche conteigneth* Lengthe, **Depeneffe.** bredth, and depeneffe. *By* Depe= neffe *I vnderftand, not as the common fort doth, the holowneffe of any thing, as of a well, a diche, a potte, and fuche like, but I meane the maffie thickneffe of*

GEOMETRICALL.

*of any bodie, as in exaumple of a potte : the depeneſſe is after
the common name, the ſpace from his brimme to his bottome.
But as I take it here, the depeneſſe of his bodie is his thickneſſe
in the ſides, whiche is an other thyng cleane different from the
depeneſſe of his holownes, that the common people meaneth.*

Now all bodies haue platte formes for their boundes, ſo
in a dye (*whiche is called a* cubike *bodie*) *by* geomdtricians,   Cubicke.
and an afhler of maſons, there are .vi. ſides, whiche are .vi.   Aſheler.
platte formes, and are the boundes of the dye.

But in a Globe, (*whiche is a bodie rounde as a bowle*)   A globe.
there is but one platte forme, and one bounde, and theſe are the
exaumples of them bothe.

A dye or aſhler   A globe.   *But becauſe you   A bounde.
                                         ſhall   not   muſe
                                         what I dooe call
                                         a bound, I mean
                                         therby   a generall
                                         name,   betokening
                                         the beginning, end
                                         and ſide, of any
                                         forme.*

A forme, fi=   Forme.
gure, or ſhape *is that thyng that is incloſed within one*   Fygure.
*bond or manie bondes, ſo that you vnderſtand that ſhape, that
the eye doth diſcerne, and not the ſubſtance of the bodie.*

Of figures *there be manie ſortes, for either thei be made of
prickes, lines or platte formes.* Notwithſtandyng *to ſpeake
properlie, a figure is euer made by platte formes, and not of
bare lines vncloſed, neither yet of prickes.*

Yet *for the lighter forme of teachyng, it ſhall not be vnſemely
to call all ſuche ſhapes, formes and figures, whiche ỹ eye maie
diſcerne diſtinctly.*

And firſt *to begin with prickes, there maie be made
diuerſe formes of them, as partely here doeth folowe.*

A.iii.   Trian=

DEFINITIONS

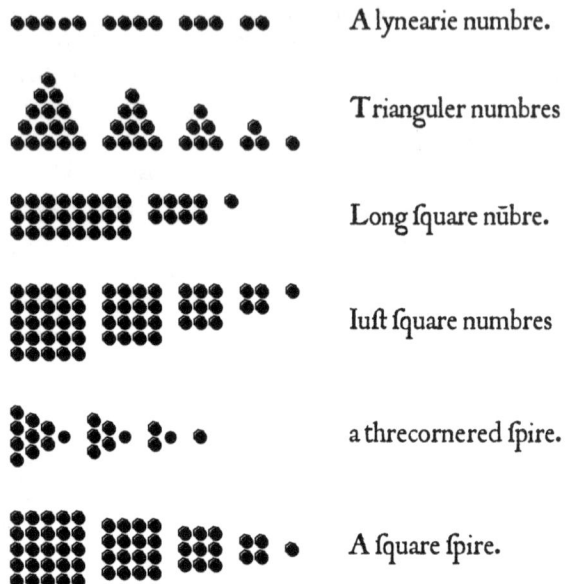

A lynearie numbre.

Trianguler numbres

Long square nūbre.

Iuſt ſquare numbres

a threcornered ſpire.

A ſquare ſpire.

    And ſo maie there be infinite formes more, whiche I omitte for this time, cōſidering that their knowledg appertaineth more to Arithmetike figurall, than to Geometrie.

    But yet one name of a pricke, whiche he taketh rather of his place then of his fourme, maie I not ouerpaſſe. And that is, when a pricke ſtandeth in the middell of a circle (as no circle can be made by cōpaſſe without it) then is it called a centre. *And thereof doe maſons, and other worke menne call that*

A centre   *patron, a* centre, *whereby thei drawe the lines, for iuſt he= wyng of ſtones for arches, vaultes, and chimneies, becauſe the chefe vſe of that patron is wrought by findyng that pricke or centre, from whiche all the lynes are drawen, as in the thirde booke it doeth appere.*

    Lynes make diuerſe figures alſo, though properly thei maie not be called figures, as I ſaid before (vnles the lines do cloſe) but onely for eaſie maner of teachyng, all ſhall be called fi= gures,

GEOMETRICALL.

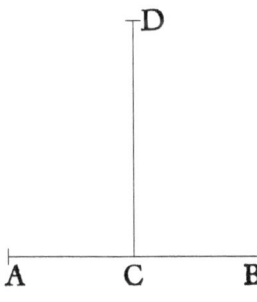

gures, that the eye can difcerne, of whiche this is one, when one line lyeth flatte (whiche is named the ground line) and an other commeth downe on it, and is cal= led a perpendiculer or plūme lyne, as in this example you may fee. Where A. B. is the grounde line, and C. D. the plumbe line.

A grounde line.
A perpen= dicular.
A plume lyne.

And likewaies in this figure there are three lines, the grounde lyne which is A. B. the plumme line that is A. C. and the bias line, whiche goeth from the one of thē to the other, and lieth againſt the right corner in fuch a figure whi= che is here C. B.
But confideryng that I ſhall haue occaſion to declare fundry figures anon, I will firſt ſhew fome certain varietees of lines that cloſe no figures, but are bare ly= nes, and of the other lines will I make mencion in the de= fcription of the figures.

Paralleles, or gemowe lynes be fuche lines as be drawen foorth ſtill in one diſtaunce, and are no nerer in one place then in an other, for and if they be nerer at one ende then at the other, then are they no paralleles, but maie bee called bought lynes, and loe here ex= aumples of them bothe.

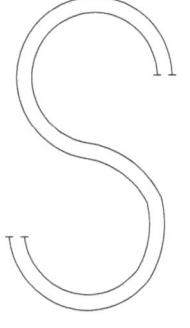

tortoufe paralleles.

Parallelys Gemowe lynes.

pa=

DEFINITIONS

I haue added alſo paralleles tortu=ouſe, *whiche bowe cõtrarie waies with their two endes : and* paralleles circu=lar, *whiche be lyke vnperfecte compaſ=ſes : for if they bee whole circles, then are they called cõ=*centrikes, *that is to ſaie, circles drawē on one centre.*

Concen=trikes.

parallelis.

parallelis circular.

Concen=trikes.

bought lines

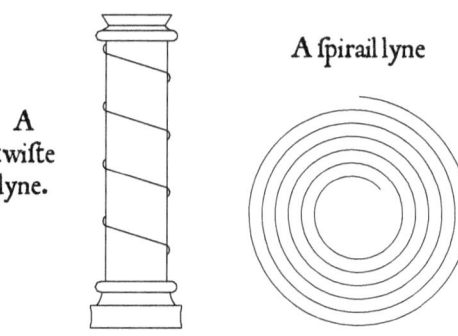

Here might I note the error of good Albert Durer, *which affirmeth that no perpendicular lines can be paralleles. which errour doeth ſpring partlie of ouerſight of the difference of a ſtreight line, and partlie of miſtakyng certain principles geo=metrical, which al I wil let paſſe vntil an other tyme, and wil not blame him, which hath deſerued worthyly infinite praiſe.*

A twine line.
A ſpirall line.
A worme line.

And to returne to my matter. an other faſhioned line is there, which is named a twine or twiſt line, and it goeth as a wreyth about ſome other bodie. And an other ſorte of lines is there, that is called a ſpirall line, or a worm line, *whiche repre=ſenteth an apparant forme of many circles, where there is not one indede : of theſe .ij. kindes of lines, theſe be examples.*

A
twiſte
lyne.

A ſpirail lyne

DEFINITION.

A touche lyne, *is a line that runneth along by the edge*   Touch line.
of a circle, *onely touching it, but doth
not croſſe the circumference of it, as in
this exaumple you maie ſee.*

*And when that a line doth croſſe the
edg of the circle, thē is it called* a cord,   A corde.
*as you ſhall ſee anon in the ſpeakynge
of circles.*

*In the meane ſeaſon muſt I not omit
to declare what angles bee called* matche corners, *that is to*   Matche
ſaie, ſuche as ſtande directly one againſt the other, when twoo   corners
*lines be drawen acroſſe, as here
appereth.*

Where A. *and* B. *are matche cor
ners, ſo are* C. *and* D. *but not* A.
and C. *nother* D. *and* A.

*Nowe will I beginne to ſpeak
of figures, that be properly ſo call
ed, of whiche all be made of di
uerſe lines, except onely a circle,
an egge forme, and a tunne forme,
which .iij. haue no angle, and haue
but one line for their bounde, and an eye fourme whiche is
made of one lyne, and hath an angle onely.*

A circle *is a figure made and encloſed with one line, and hath*   A circle.
*in the middell of it a pricke or centre, from whiche all the
lines that be drawen to the circumfernece are equall all in
length, as here you ſee.*

*And the line that encloſeth the
whole compaſſe, is called the circum*   Circumfe=
ference.   rence.

And all the lines that bee drawen   A diameter
*croſſe the circle, and goe by the centre,
are named* diameters, *whoſe halfe, I
meane from the center to the circum=*

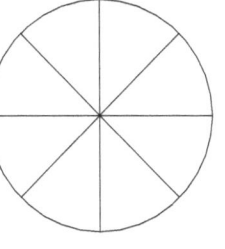

B   ference

DEFINITIONS

Semidia=
meter.

ference *any waie, is called the* femidiameter, *or* halfe diameter.

*But and if the line goe croſs the cir=
cle, and paſſe beſide the centre, then is*
A cord or a
ſtring lyne.
it called a corde, *or a* ſtryng line, *as I ſaid before, and as this exaumple ſheweth : where* A. *is the corde.*

And the compaſſed line that aunſwe=
An arch line. *reth to it, is called* an arche lyne, *or*
A bowline.   a bowe lyne, *whiche here is marked with* B. *and the diameter with* C.

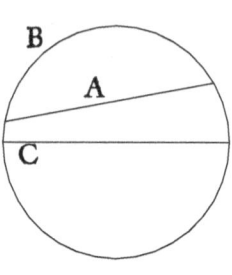

But and if that part be ſeparate from the reſt of the circle (as in this exāple you ſee) then ar both
A cantle.   *partes called cā=
telles,   the   one*   the greatter cantle, *as* E. *and the other the* leſſer cantle,
A femye   *as* D. *And if it be parted iuſte by the centre (as you ſee in* F.)
circle   *then is it called a* femicircle, *or* halfe compaſſe.

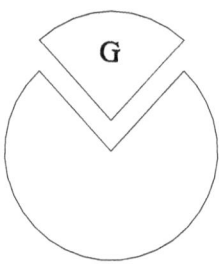

Sometimes it happeneth that a cantle is cutte out with two lynes drawen from the centre to the circumference (as G. *is)*
A nooke
cantle.
and then maie it be called a nooke can=
tle, *and if it be nat parted from the reſte of the circle (as you ſee in* H.) *then is it*
A nooke.   called a nooke *plainely without any addicion. And the compaſſed lyne in it is called an* arche lyne, *as the exaum=
ple here doeth ſhewe.*

Now

GEOMETRICALL.

An arche.

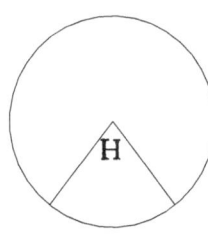

Nowe haue you heard as touchyng circles meetely sufficient instruction, so that it should seme nedeles to speake any more of figures in that kynde, saue that there doeth yet remaine ij. formes of an imperfecte circle, for it is lyke a circle that were brused, and thereby did runne out ende long one waie, whiche forme Geometricians dooe call an

egge forme, *becaufe it doeth represent the figure and shape of an egge duely proportioned (as this figure sheweth) hauyng the*

An egge forme.

An egge forme.

A tunne forme.

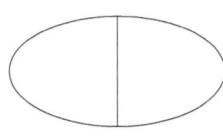

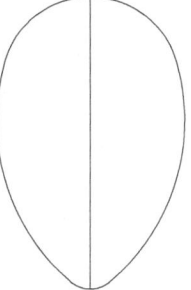

one ende greater then the other.

For if it be lyke the figure of a circle pressed in length, and bothe endes lyke bygge, then is it called a tunne forme, *or* barrell forme, *the right makyng of whiche figures, I wyll declare hereafter in the thirde booke.*

An other forme there is, whiche you maie call a nutte forme, and is made of one lyne muche lyke an egge forme, saue that it hath a sharpe angle.

And it chaunceth sometyme that there is a right line drawen crosse these figures, and that is called an axelyne, *or ax= tre. Howe be it properly that line that is called an* axtre, *whiche gooeth thoroughe the myddell of a* Globe, *for as a diameter is in a circle, so is an axe lyne or axtre in a* Globe,

A tunne or barrall form

An axtre or axe lyne.

DEFINITIONS

that lyne that goeth from fide to fyde, and paffeth by the mid
dell of it. And the two poyntes that fuche a lyne maketh in
the vtter bounde or platte of the globe, are named polis, w[ch]
you may call aptly in englyfh, tourne pointes : of whiche I
do more largely intreate, in the booke that I haue written of
the vfe of the globe.

But to returne to the diuerfityes of figures that remayne
vndeclared, the moft fimple of them ar fuch ones as be made
but of two lynes, as are the cantle of a circle, and the halfe
circle, of which I haue fpoken all ready. Likewyfe the halfe
of an egge forme, the cantle of an egge forme, the halfe
of a tunne fourme, and the cantle of a tunne fourme,
and befyde thefe a figure moche like to a tunne fourme, faue
that it is fharp couered at both the en=
des, and therfore doth confift of twoo
lynes, where a tunne forme is made
An yey fo= of one lyne, and that figure is named
urme.   an yey fourme.

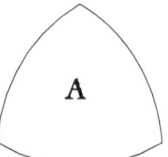

A triangle.   The nexte kynd of figures are thofe
that be made of .iij. lynes other be all right lynes, all crooked
lynes, other fome right and fome crooked. But what fourme
fo euer they be of, they are named generally triangles for a tri
angle is nothinge elfe to fay, but a figure of three corners.
And thys is a generall rule, looke how
many lynes any figure hath, fo mannye
corners it hath alfo, yf it bee a platte
forme, and not a bodye. For a bodye
hath dyuers lynes metyng fometime
in one corner.

Now to geue you example of tri
angles, there is one whiche is all of cro=
ked lynes, and may be taken fur a por=
tio of a globe as the figur marked w A.

An other hath two compaffed lines
and one right lyne, and is as the porti
on of halfe a globe, example of B.

An other hatht but one compaffed

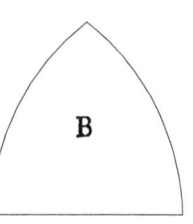

                                                        lyne

GEOMETRICALL.

lyne, and is the quarter of a circle, named a quadrate, and the ryght lynes make a right corner, as you se in C. Otherleſſe then it as you ſe D, whoſe right lines make a ſharpe corner, or greater then a quadrate, as is F, and then the right lynes of it do make a blunt corner.

Alſo ſome triangles haue all righte lynes and they be diſtinɛ̃ted in ſonder by their an= gles, or corners, for other their corners bee all ſharpe, as you ſee in the figure, E. other ij. ſharpe and one right ſquare, as is the figure G other ij. ſharp and one blunt as in the figure H.

There is alſo an other diſtinɛ̃tion of the names of triangles, according to their ſides, whiche other be all equall as in the figure E, and that the Greekes doo call Iſopleuron, and Latine men æquilaterum : and in engliſh it may be called a threlike triangle, other els two ſydes bee e= quall and the thyrd vnequall, which the Greekes call Iſoſceles, the La= tine men æquicurio, and in engliſh tweyleke may they be called, as in G, H, and K. For, they may be of iij. kinds that is to ſay, with one ſquare angle, as is G, or with a blunte corner as H, or with all in ſharpe korners, as you ſee in K.

Furthermore it may be y̓ they haue neuer a one ſyde equall to an other, and they be in iij kyndes alſo diſtinɛ̃t lyke the twilekes, as you maye perceaue by theſe examples .M. N, and O where M hath a right angle, N, A, blunte angle, and O, all ſharpe angles theſe the Greekes and latine men do

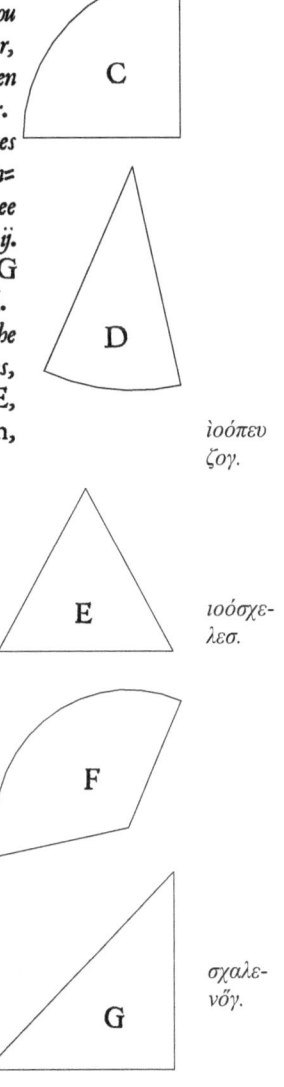

ἰσόπευ ζογ.

ἰσόσχε- λεσ.

σχαλε- νόγ.

DEFINITIONS

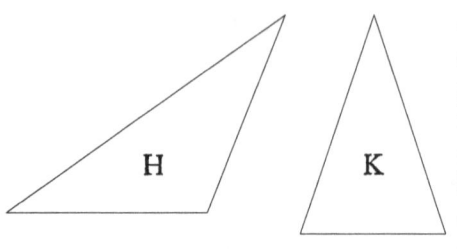

cal fcalena
and in en=
glifhe theye
may be cal=
led nouele
kes, *for thei*
haue no fide
equall,   or

like lõg, to ani other in the fame figur.
Here it is to be noted, that in a triãgle
al the angles bee called innerãgles
                              except ani fide
                              bee drawenne
                              forth in leng=
                              the, for then
                              is that fourthe

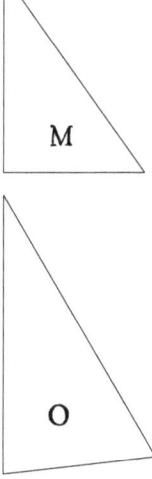

M

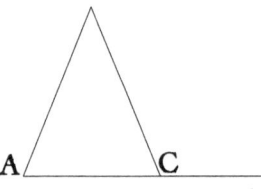

corner caled an
vtter corner,
as in this exãple
becaufe A, B, *is drawen in length, ther*
fore the ãgle C, *is called an vtter ãgle.*

And thus haue I done with triãguled
Quadrangle *figures, and nowe foloweth* quadran
gles, *which are figures of iiij. corners*
and of iiij. lines alfo, of whiche there

O

                      be diuers kindes, but chiefe
                      ly v. that is to fay, a fquare
A fquare   quadrate, *whofe fides bee*
quadrate.   *all equall, and al the angles*
                      fquare, as you fe here in this
                      figure Q. *The fecond kind*

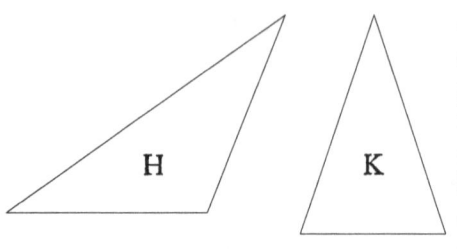

A$_$ C   B

A longe   *is called a long fquare, whofe foure cor*
fquare.   *ners be all fquare, but the fides are not*
                      equall eche to other, yet is euery fide
                      equall to that other that is againft it, as
                      you maye perceaue in this figure R.
                                                        The

Q

GEOMETRICALL.

The thyrd kind is called lofenges *or* diamondes, *whofe fides bee all e=quall, but it hath neuer a fquare corner, for two of them be fharpe, and the other two be blunt, as appeareth in,* S.

The iiij. *forte are like vnto lofenges, faue that they are longer one waye, and their fides be not equal, yet ther corners are like the corners of a lofing, and ther fore ar they names* lofengelike *or di=*amodlike, *whofe figur is noted with* **T**. *Here fhal you marke that al thofe fqua=res which haue their fides al equal, may be called alfo for eafy vnderftandinge,* likefides, *as* **Q**. *and* S. *and thofe that haue only the contrary fydes equal, as* R. *and* **T**. *haue, thofe wyll I call* like=iammys, *for a difference.*

The fift forte doth containe all other fafhions of foure cornered fi=gurs, and ar called of the Grekes tra pezia, *of Latin* me menfulæ *and of* Arabitians, helme=ariphe, *they may be called in englifhe* borde formes, *they haue no fyde e=quall to an other as thefe examples fhew, neither keepe they any rate in their corners, and therfore are they counted* vn=ruled formes, *and the other foure kindes only are counted* ruled formes, *in the kynde of quadrangles. Of thefe vnru led formes ther is no numbre, they are fo mannye and fo dy=uers, yet by arte they may be changed into other kindes of fy=gures, and therby be brought to meafure and proportion, as in the thirtene conclufion is partly taught, but more plainly in my booke of meafuring you may fee it.*

A lofenge.
A diamōde

A lofenge lyke.

Borde for mes.

And

DEFINITIONS

And nowe we make an eande of the dyuers kyndes of figures, there dothe folowe now figures of .v. sydes, other v. corners, which we may call cink=angles, *whose sydes partlye are all e=quall as in* A, *and those are counted* ruled cinkeangles *and partlye vne=quall as in,* B *and they are called* vnru led.

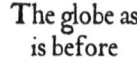

Likewyse shall you iudge of fisean=gles, *which haue sixe corners,* septan gles, *whiche haue seuen angles, and so forth, for as mannye numbres as there maye be of sydes and angles, so manye di=uers kindes be there of figures, vnto which yow shall geue names according to the numbre of their sides and angles, of whiche for this tyme I wuill make an ende, and wyll sette forthe on example of a syseangle, which I had almost for=gotten, and that is it, whose vse com=meth often in* Geometry, *and is called a* squire, *is made of two long squares ioy ned togither, as this example sheweth.*

And thus I make an eand to speak of platte formes, and will briefelye saye somewhat touching the figures of bodeis *which partly haue one platte forme for their bound, and ÿ iust roůd as a* globe *hath, or ended long as in an* egge, *and a* tunne fourme, *whose pictures are these.*

Howe be it you must marke that I meane not the very fi=gure of a tunne, when I saye tunne form, but a figure like a tunne, for a tune forme,

The globe as
is before

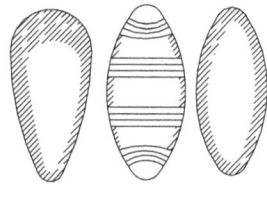

hath

GEOMETRICALL.

*hath but one plat forme, and therfore mu∫t needs be round at
the endes, where as a* tunne *hath thre platte formes, and is
flatte at eche end, as partly the∫e pi∂ures do ∫hewe.*

Bodies of two plattes *are other cantles or halues of
tho∫e other bodies, that haue but one platte forme, or els
they are lyke in foorme to two ∫uch cantles ioyned togither
as this* A *doth partly eppre∫∫e : or els
it is called a* rounde ∫pire, *or* ∫ti=
ple fourme, *as in this figure is ∫ome
what expre∫∫ed.*

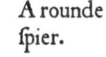

A rounde
∫pier.

Nowe *of three plattes there are
made certain figures of bodyes, as the
cantels and halues of all bodyes that
haue but j. plattys and al∫o the hal=
ues of halfe globys and canteles of
a globe. Lykewy∫e a rounde piller,
and a ∫pyre made of a rounde ∫pyre,
∫lytte in ij. partes long ways.*

But *as the∫e formes be harde to be iudged by their py∂urs,
∫o I doe entende to pa∫∫e them ouer with a great number of
other formes of bodyes, which afterwarde ∫hall be ∫et forth
in the boke of* Per∫pe∂iue, *bicau∫e that without per∫pe∂iue
knowledge, it is not ea∫y to iudge truly the formes of them in
flatte prota∂ure.*

*And thus I make an ende for this tyme, of the defi=
nitions Geometricall, appertayning to this
parte of pra∂i∫e, and the re∫t wil
I pro∫ecute as cau∫e ∫hall
∫erue.*

C

THE PRACTIKE WORKINGE OF
ſondry concluſions Geometrical.

THE FYRST CONCLVSION.
*To make a threlike triangle or any lyne
meaſurable.*

AKE THE IVSTE
*lēgth of the lyne with your cōpaſſe,
and ſtay the one foot of the compas
in one of the endes of that line, tur
ning the other vp or down at your
will, drawyng the arche of a circle
againſt the
midle of the*
line, and doo likewiſe with the ſame
cōpaſſe vnaltered, at the other end of
the line and wher theſe ij. croked ly=
nes doth croſſe, frome thence drawe a
lyne to ech end of your firſt line, and
there ſhall appear a threlike triangle
drawen on that line.

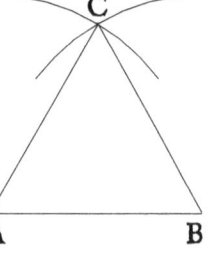

Example.

A.B. *is the firſt line, on which I wold
make the threlike triangle, therfore I
open the compaſſe as wyde as that line
is long, and draw two arch lines that
mete in* C, *then from* C *I draw ij other
lines one to* A, *another to* B, *and than
I haue my purpoſe.*

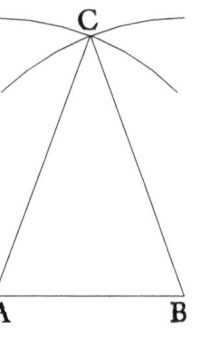

THE II. CONCLVSION.
*If you wil make a twileke or
a nouelike triangle on ani cer
taine line.*

*Conſider fyrſt the length that yow will haue the other ſi=
des to containe, and to that length open your compaſſe, and
then*

CONCLVSIONS GEO.

then worke as you did in the threleke triangle, remembryng
this, that in a nouelike triangle you muſt take ij. lengthes be=
ſyde the fyrſte lyne, and draw an arche lyne with one of the̅
at the one ende, and with the other at
the other end, the exa̅ple is as in the o=
ther before.

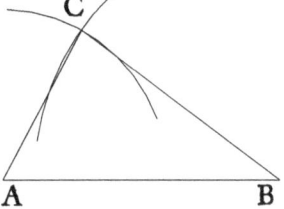

THE III. CONCL.

*To diuide an angle of right
lines into ij. equal partes.*

Firſt open your compaſſe as largely as you can, ſo thaſt it do
not excede the length of the ſhorteſt line ẙ incloſeth the an=
gle. Then ſet one foote of the compaſſe in the verye point of
the angle, and with the other fote draw a compaſſed arch fro̅
the one lyne of the angle to the other,
that arch ſhall you deuide in halfe, and
the̅ draw a line fro̅ the a̅gle to ẙ mid=
dle of ẙ arch, and ſo ẙ angle is diuided
into ij. equall partes.   Example.
Let the tria̅gle be A.B.C, the̅ ſet I one
foot of ẙ copaſſe in B, and with the o=
ther I draw ẙ arch D.E, which I part
into ij. equall parts in F, and the̅ draw
a line fro̅ B, to F, & ſo I haue mine inte̅t

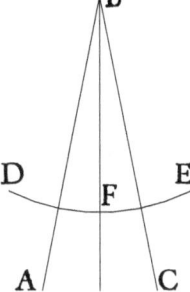

THE IIII. CONCL.

*To deuide any meaſurable
line into ij. equall partes.*

Open your compaſſe to the iuſt le̅gth of
ẙ line. And the̅ ſet one foote ſteddely at
the one ende of the line, & ẘ the other
fote draw an arch of a circle againſt ẙ
midle of the line, both ouer it, and al=
ſo vnder it, then doo lykewaiſe
                    C.ij.   at the

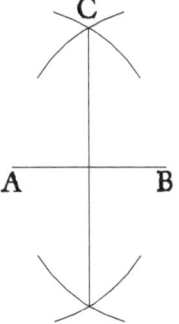

CONCLVSIONS

at the other ende of the line. And marke where thofe arche lines do meet croffewaies, and betwene thofe ij. pricks draw a line, and it fhall cut the firft line in two equall portions.

Example.

The lyne is A. B. *accordyng to which I open the compaffe and make .iiij. arche lines, whiche meete in* C. *and* D, *then drawe I a lyne from* C, *fo haue I my purpofe.*

This conclufion ferueth for makyng of quadrates and fqui= res, befide many other commodities, howebeit it maye bee don more readylye by this conclufion that foloweth nexte.

THE FIFT CONCLVSION.

To make a plumme line or any pricke that you will in any right lyne appointed.

Open youre compas fo that it be not wyder then from the pricke appoynted in the line to the fhorteft ende of the line, but rather fhorter. Then fette the one foote of the compaffe in the firfte pricke appointed, and with the other fote marke ij. other prickes, one of eche fyde of that fyrfte, afterwarde open your compaffe to the wydenes of thofe ij. new prickes,

and draw from them ij. arch lynes, as you did in the fyrft conclufion, for making of a threlyke triãgle. then if you do mark their croffing, and from it drawe a line to your fyrfte pricke, it fhall bee a iuft plum lyne on that place.

Example

The lyne is A.B. *the prick on whiche I fhoulde make the plumme lyne, is* C. *then open I the compaffe as wyde as* A.C. *and fette one foote in* C. *and with the other doo I marke out* C.A. *and* C.B. *then open I the compaffe as wide as* A.B. *and make ij. arch lines which do croffe in* D, *and fo haue I doone.*

Howe bee it, it happeneth fo fommetymes, that the
pricke

GEOMETRICALL.

pricke on whiche you would make the perpendicular or plum line, is so nere the eand of your line, that you can not extende any notable length from it to thone end of the line, and if so be it then that you maie not drawe your line lenger frō that end, then doth this conclusion require a newe ayde, for the last de= uise will not serue. In suche case therfore shall you dooe thus : If your line be of any notable length, deuide it into fiue partes. And if it be not so long that it maie yelde fiue notable partes, then make an other line at will, and parte it into fiue equall portiōs : so that thre of those partes maie be found in your line. Then open your compasse as wide as thre of these fiue measures be, and sette the one foote of the compas in the pricke, where you would haue the plumme line to lighte (which I call the first pricke,) and with the other foote drawe an arche line righte ouer the pricke, as you can ayme it : then open youre compas as wide as all fiue measures be, and set the one foote in the fourth pricke, and with the other foote draw an other arch line crosse the first, and where thei two do crosse, thense draw a line to the poinct where you woulde haue the perpendicular line to light, and you haue doone.

Example.

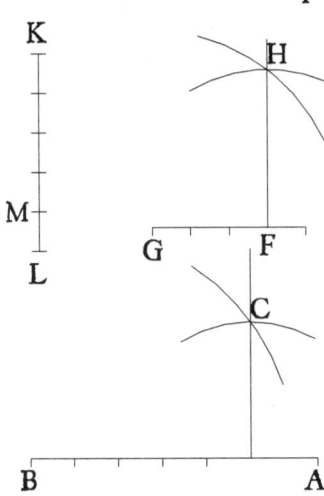

The line is **A.B.** and A. is the pricke on whi= che the perpendicular line must light. Therfore I deuide **A.B.** into fiue partes equall, then do I open the compas to the widenesse of three par= tes (that is **A.D.**) and set one foote staie in **A.** and with the other I make an arche line in **C.** Afterwarde I open the compas as wide as **A.B.**

C.iij.   (that

DEFINITIONS

(that is as wide as all fiue partes) and set one foote in the .iiij.
pricke, which is E, drawyng an arch line with the other foote
in C. also. Then do I draw thence a line vnto A, and so haue
I doone. But and if the line be to shorte to be parted into fiue
partes, I shall deuide it into .iij. partes only, as you see the line
F.G. and then make D. an other line (as is K.L.) whiche I
deuide into .v. suche diuisions, as F.G. containeth .iij, then o=
pen I the compaas as wide as .iiij. partes (whiche is K.M.)
and so set I one foote of the compas in F, and with the other I
drawe an arch lyne toward H, then open I the cōpas as wide
as K.L. (that is all .v. partes) and set one foote in G, (that is the
iij. pricke) and with the other I draw an arch line toward H.
also : and where those .ij. arch lines do crosse (which is by H.)
thence draw I a line vnto F, and that maketh a very plumbe
line to F.G. as my desire was. The maner of workyng of this
conclusion, is like to the second conclusion , but the reason of it
doth depēd of the .xlvi. proposiciō of y̆ first boke of Euclide.
An other waie yet. set one foote of the compas in the prick, on
whiche you would haue the plumbe line to light, and stretche
forthe thother foote toward the longest end of the line, as wide
as you can for the length of the line, and so draw a quarter of a
compas or more, then without stirryng of the compas, set one
foote of it in the same line, where as the circular line did begin,
and extend thother in the circular line, settyng a marke where
it doth light, then take half that quan
titie more there vnto, and by that
prick that endeth the last part, draw
a line to the pricke assigned, and it
shall be a perpedicular.

Example.

A.B. is the line appointed, to whi=
che I must make a perpedicular line
to light in the pricke assigned, which
is A. Therfore doo I set one foote of
the compas in A, and extend the o=
ther vnto D. makyng a part of a cir=

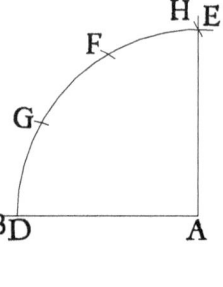

cle,

GEOMETRICALL.

cle, more then a quarter, that is D.E. *Then do I set one foote of the compas vnaltered in* D, *and stretch the other in the cir=cular line, and it doth light in* F, *this space betwene* D. *and* F. *I deuide into halfe in the pricke* G, *whiche halfe I take with the compas, and set it beyond* F. *vnto* H, *and therfore is* H. *the point, by whiche the perpendicular line must be drawen, so say I that the line* H.A. *is a plumbe line to* A.B. *as the conclu=sion would.*

THE VI. CONCLVSION.

To drawe a streight line from any pricke that is not in a line, and to make it perpendi=cular to an other line.

Open your compas so wide that it may extend somewhat far=ther, thē from the prick to the line, then sette the one foote of the compas in the pricke, and with the other shall you draw a cōpassed line, that shall crosse that other first line in .ij. places Now if you deuide that arch line into .ij. equall partes, and from the middell pricke ther=of vnto the prick without the line you drawe a streight line, it shalbe a plumbe line to that firste lyne, accordyng to the conclusion. Example.

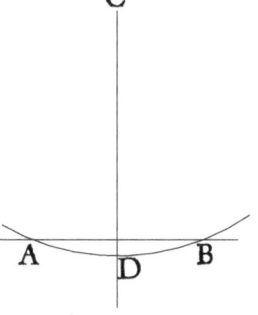

C. *is the appointed pricke, from whiche vnto the line* A.B. *I must draw a perpēdicular. Therfore I open the cōpas so wide, that it may haue one foote in* C, *and thother to reach ouer the line, and with ẏ foote I draw an arch line as you see, betwene* A. *and* B, *which arch line I deuide in the middell in the point* D. *Then drawe I a line from* C. *to* D, *and it is perpendicu=lar to the line* A.B. *accordyng as my desire was.*

The

CONCLVSIONS

THE VII. CONCLVSION.

To make a plumbe lyne or any porcion of a circle, and that on the vtter or inner bughte.

Mark firſt the prick where ẏ plūbe line ſhal lyght : and prick out on ech ſide of it .ij. other poinĉtes equally diſtant from that firſt pricke. Then ſet the one foote of the cōpas in one of thoſe ſide prickes, and the other foote in the other ſide pricke, and firſt moue one of the feete and drawe an arche line ouer the middell pricke, then ſet the compas ſteddie with the one foote in the other ſide pricke, and with the other foote drawe an o=ther arche line, that ſhall cut that firſt arche, and from the ve=ry poinĉte of their meetyng, drawe a right line vnto the firſte pricke, where you do minde that the plumbe line ſhall lyghte. And ſo haue you performed thintent of this concluſion.

Example.

The arche of the circle on whiche I would erect a plumbe line, is A.B.C. and B. is the pricke where I would haue the

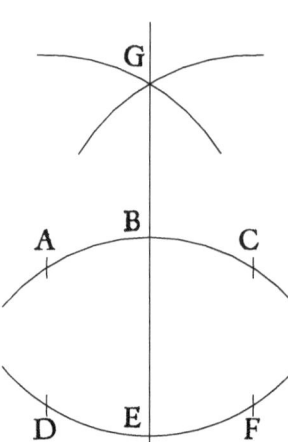

plumbe line to light. Ther=fore I meate out two equall diſtaunces on eche ſide of that pricke B. and they are A.C. Then open I the com=pas as wide as A.C. and ſet=tyng one of the feete in A. with the other I drawe an arch line which goeth by G. Likewaies I ſet one foote of the compas ſteddily in C. and with the other I drawe an arche line, goyng by G. alſo. Now conſideryng that G. is the pricke of their meetyng, it ſhall be alſo the poinĉt from whiche I muſt drawe the plūbe line. Then draw I a rightline from G. to B. and ſo haue mine intent. Now as A.B.C. hath a plumbe line erected on his

vtter

GEOMETRICALL.

vtter bought, ſo may I erect a plumbe line on the inner bught of D.E.F, doynge with it as I did with the other, that is to ſaye, fyrſte ſettyng forthe the pricke where the plumbe line ſhall light, which is E, and then markyng one other on eche ſyde, as are D. and F. And then proceding as I dyd in the ex= ample before.

THE VIII. CONCLVSYON.

How to deuide the arche of a circle into two equall partes, without meaſuring the arche.

Deuide the corde of that line into ij. equall portions, and then from the middle prycke erecte a plumbe line, and it ſhal parte that arche in the middle.
        Example.
The arch to be diuided ys A.D.C, *the corde is* A,B,C, *this corde is diuided in the middle with* B, *from which prick if I erecte a plum line as* B.D, *thē will it diuide the arch in the middle, that is to ſay, in* D.

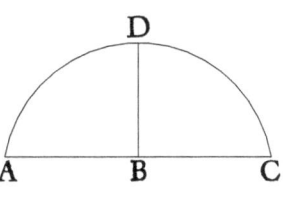

THE IX. CONCLVSION.

To do the ſame thynge other wiſe. And for ſhortenes of worke, if you wyl make a plumbe line without much labour, you may do it with your ſquyre, ſo that it be iuſtly made, for yf you applye the edge of the ſquyre to the line in which the prick is, and foreſee the very corner of the ſquyre doo touche the pricke. And than frome that corner if you drawe a lyne by the other edge of the ſquyre, yt will be a perpendicular to the former line.

                    D   *Example*

CONCLVSIONS

Example.

A.B. *is the line, on which I wold make the plumme line, or perpendicular. And there= fore I marke the prick from which the plumbe lyne mufte rife, which here is* C. *Then do I fette one edg of my fquyre* (*that is* B.C.) *to the line* A.B, *fo that the corner of the fquyre do touche* C. *iuftly. And from* C. *I drawe a line by the other edge of the fquire,* (*which is* C.D.) *And fo haue I made the plumme line* D.C, *which I fought for.*

THE X. CONCLVSION.

Howe to do the fame thinge an other way yet

If fo be it that you haue an arche of fuche greatnes, that your fquyre wyll not fuffice therto, as the arche of a brydge or of a houfe or window, then may you do this. Mete vnderneth the arch where y̆ midle of his cord wyl be, and ther fet a mark Then take a long line with a plummet, and holde the line in fuche a place of the arch, that the plummet do hang iuftely o= uer the middle of the corde, that you didde diuide before, and then the line doth fhewe you the middle of the arche.

Example.

The arch is A.D.B, *of which I trye the midle thus. I draw a corde from one fyde to the other* (*as here is* A.B,) *which I diuide inthe middle in* C. *Thē take I a line with a plum met* (*that is* D.E,) *and fo hold I the line that the plummet* E, *dooth hange ouer* C, *And then*

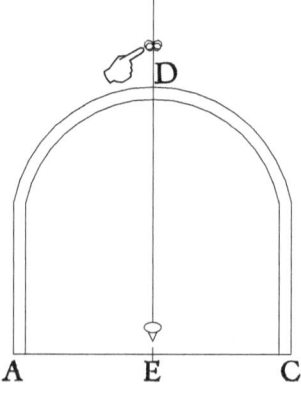

GEOMETRICALL.

then I fay that D. *is the middle of the arche. And to thentent that my plummet fhall point the more iuftely, I doo make it fharpe at the nether ende, and fo may I truft this woorke for certain.*

THE XI. CONCLVSION.

When any line is appointed and without it a pricke, whereby a parallel muft be drawen howe you fhall doo it,

Take the iufte meafure beetwene the line and the pricke, accordinge to which you fhal open your compaffe. Thē pitch one foote of your compaffe at the one ende of the line, and with the other foote draw a bowe line right ouer the pytche of the compaffe, lykewife doo at the other ende of the lyne, then draw a line that fhall touche the vttermofte edge of bothe thofe bowe lines, and it will bee a true parallele to the fyrfte lyne appointed.

Example.

A.B, *is the line vnto which I muft draw an other gemow line, which mufte paffe by the prick* C, *firft I meate with my compaffe the fmalleft diftance that is from* C. *to the line, and that is* C.F, *wherfore ftaying the compaffe at that diftaunce, I fette the one foote in* A, *and with the other foot I make a bowe lyne, which is* D, *thē like wife fet I the one foote of the compaffe in* B, *and with the o= ther I make the fecond bow line, which is* E. *and then draw I a line, fo that it toucheth the vttermoft edge of bothe thefe bowe lines, and that lyne paffeth by the pricke* C, *and is a ge= mowe line to* A.B, *as my fekyng was.*

D.ij.   The

CONCLVSIONS

THE XI. CONCLVSION.

To make a triangle of any .iij. lines, fo that the lines be fuche, that any .ij. of them be lon= ger then the thirde. For this rule is generall, that any two fides of euerie triangle taken to= gether, are longer then the other fide that re= maineth.

If you do remember the firft and feconde conclufions, then is there no difficultie in this, for it is in maner the fame worke. Firft cõfider the .iij. lines that you muft take, and fet one of thē for the ground line, then worke with the other .ij. lines as you did in the firft and fecond conclufions.

Example.

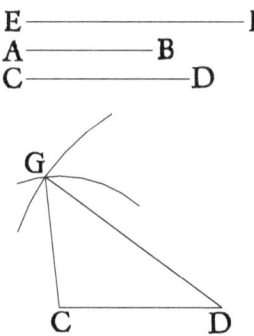

I haue .iij. lynes .A.B. and C.D. and E.F. of which I put .C.D. for my ground line, then with my compas I take the length of .A.B. and fet the one foote of my compas in C, and draw an arch line with the other foote. Likewaies I take the lēgth of E.F, and fet one foote in D, and with the other foote I make an arch line croffe the other arche, and the pricke of their me= tyng (which is G.) fhall be the thirde corner of the triangle, for in all fuche kyndes of woorkynge to make a tryangle, if you haue one line drawen, there remayneth nothyng els but to fynde where the pitche of the thirde corner fhall bee, for two of them muft needes be at the two eandes of the lyne that is drawen.

The

GEOMETRICALL.

THE XIII CONCLVSION.

If you haue a line appointed, and a pointe in it limited, howe you maye make on it a righte lined angle, equall to an other right lined angle, all ready assigned.

Fyrst draw a line against the corner assigned, and so is it a triangle, then take heede to the line and the pointe in it assigned, and consider if that line from the pricke to this end bee as long as any of the sides that make the triangle assigned, and if it bee longe inoughe, then prick out there the length of one of the lines, and then woorke with the other two lines, accordinge to the laste conclusion, makynge a triangle of thre like lynes to that assigned triangle. If it bee not longe inoughe, thenne lengthen it fyrste, and afterwarde doo as I haue sayde beefore.

Example.

Lette the angle appoynted bee A.B.C, and the corner assigned, B. Farthermore let the lymited line bee D.G, and the pricke assigned D.

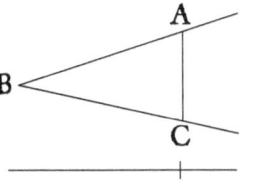

Fyrste therefore by drawinge the line A.C, I make the triangle A.B.C.

Then consideringe that D.G, is longer thanne A.B, you shall cut out a line frō D. toward G, equal to A.B, as for exāple D,F. Thē measure oute the other ij. lines and worke with thē according as the conclusion with the fyrste also and the second teacheth yow, and then haue you done.

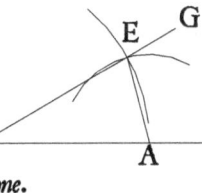

D.iij.   The

DEFINITIONS

THE XIIII. CONCLVSION.

To make a square quadrate of any righte lyne appoincted.

First make a plumbe line vnto your line appointed, whiche shall light at one of the endes of it, accordyng to the fifth conclusion, and let it be of like length as your first line is, then ope your compasse to the iuste length of one of them, and sette one foote of the compasse in the ende of the one line, and with the other foote draw an arche line, there as you thinke that the fowerth corner shall be, after that set the one foote of the same compasse vnsturred, in the eande of the other line, and drawe an other arche line crosse the first arche line, and the poincte that they do crosse in, is the pricke of the fourth corner of the square quadrate which you seke for, therfore draw a line from that pricke to the eande of eche line, and you shal therby haue made a square quadrate.

Example.

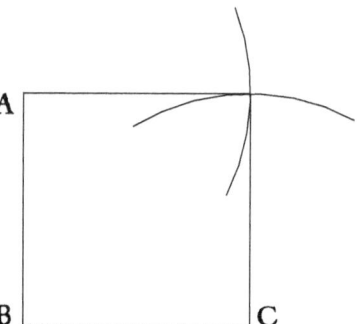

A.B. *is the line proposed, of whiche I shall make a square quadrate, therefore firste I make a plūbe line vnto it, whiche shall lighte in* A, *and that plūb line is* A.C. *then open I my compasse as wide as the length of* A.B, *or* A.C, *(for they must be bothe equall) and I set the one foote of thend in* C, *and with the other I make an arche line nigh vnto* D, *afterward I set the compas again with one foote in* B, *and with the other foote I make an arche line crosse the first arche line in* D, *and from the prick of their crossyng I draw .ij. lines, one to* B, *and an other to* C, *and so haue I made the square quadrate that I entended.*

The

THE XV. CONCLVSION.

To make a likeiāme equall to a triangle ap=
pointed, and that in a right lined āgle limited.

Firſt from one of the angles of the triangle, you ſhall drawe
a gemowe line, whiche ſhall be a parallele to that ſyde of the
triangle, on whiche you will make that likeiamme. Then on
one end of the ſide of the triangle, whiche lieth againſt the ge=
mowe lyne, you ſhall draw forth a line vnto the gemow line,
ſo that one angle that commeth of thoſe .ij. lines be like to the
angle whiche is limited vnto you. Then ſhall you deuide into
ij. equall partes that ſide of the triangle whiche beareth that
line, and from the pricke of that deuiſion, you ſhall raiſe an o=
ther line parallele to that former line, and continewe it vnto
the firſt gemowe line, and thē of thoſe .ij. laſt gemowe lynes,
and the firſt gemowe line, with the halfe ſide of the triangle,
is made a lykeiamme equall to the triangle appointed, and hath
an angle lyke to an angle limited, accordyng to the concluſion.

Example.

B.C.G, *is the tri=*
angle appointed
vnto, whiche I
muſte make an e=
quall likeiamme.
And D, *is the an=*
gle that the like=
iamme muſt haue.
Therfore firſt en=
tendyng to erecte

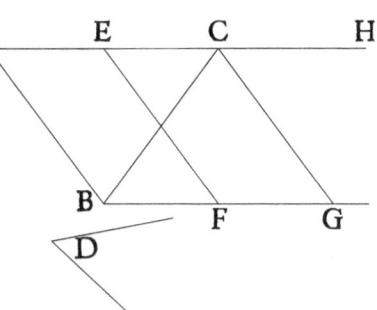

the likeiāme on the one ſide, that the ground line of the trian=
gle (whiche is B.G.) I do draw a gemow line by C, and make
it parallele to the ground line B.G, and that new gemow line
is A.H. Then do I raiſe a line from B. vnto the gemowe line,
(whiche line is A.B) and make an angle equall to D, that is the
appointed angle (accordyng as the .viij. cōcluſion teacheth, and
that angle is B.A.E. Then to procede, I doo parte in ẙ middle
the ſaid groūd line B.G, in the prick F, frō which prick I draw
                                                                        to

CONCLVSIONS

to the firſt gemowe line (A.H.) an other line that is parallele
to A.B, and that line is E.F. Now ſaie I that the likeiāme
B.A.E.F, is equall to the triangle B.C.G. And alſo that it
hath one angle (that is B.A.E. like to D. the angle that was
limitted. And ſo haue I mine intent. The profe of the equal=
nes of thoſe two figures doeth depend of the .xli. propoſition
of Euclides firſt boke, and is the .xxxi. propoſition of this ſe=
cond boke of Theoremis, whiche ſaieth, that whan a tryangle
and a likeiamme be made betwene .ij. ſelfe ſame gemow lines,
and haue their ground line of one length, then is the likeiamme
double to the triangle, wherof it foloweth, tht if .ij. ſuche fi=
gures ſo drawen differ in their ground line onely, ſo that the
ground line of the likeiamme be but halfe the ground line of
the triangle, then be thoſe .ij. figures equall, as you ſhall more at
large perceiue by the boke of Theoremis, in y̆ .xxxi. theoreme.

THE XVI. CONCLVSION.

*To make a likeiamme equall to a triangle
appoincted, accordyng to an angle limitted,
and on a line alſo aſſigned.*

In the laſt concluſion the ſides of your likeiamme wer left
to your libertie, though you had an angle appointed. Nowe
in this concluſion you are ſomwhat more reſtrained of libertie
ſith the line is limitted, which muſt be the ſide of the likeiāme.
Therfore thus ſhall you procede. Firſte accordyng to the laſte
concluſion, make a likeiamme in the angle appointed, equall
to the triangle that is aſſigned. Then with your compaſſe take
the length of your line appointed, and ſet out two lines of the
ſame length in the ſecond gemowe lines, beginnyng at the one
ſide of the likeiamme, and by thoſe two prickes ſhall you draw
another gemowe line, whiche ſhall be parallele to two ſides
of the likeiamme. Afterward ſhall you draw .ij. lines more for
the accompliſhement of your worke, whiche better ſhall be
                                                        perceiued

GEOMETRICALL.

perceaued by a ſhorte exaumple, then by a greate numbre of wordes, only without example, therefore I wyl by example ſette forth the whole worke.

Example.

Fyrſt, according to the laſt concluſion, I make the like=iamme E.F.C.G, *equal to the triangle* D, *in the appoynted angle whiche is* E. *Then take I the lengthe of the aſſigned line (which is* A.B,*) and with my compas I ſette forthe the ſame lēgth in the ij. gemow li nes* N.F. *and* H.G, *ſetting one foot in* E, *and the other in* N, *and againe ſettyng one foote in* C, *and the other in* H. *Af=terward I draw a line from* N. *to* H, *whiche is a gemow*

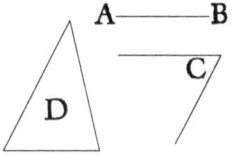

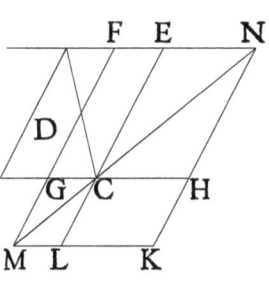

lyne, to ij. ſydes of the likeiamme. thenne drawe I a line alſo from N. *vnto* C, *and extend it vntyll it croſſe the lines,* E.L. *and* F.G, *which both muſt be drawen forth longer then the ſi=des of the likeiamme. and where that lyne doeth croſſe* F.G, *there I ſette* M. *Nowe to make an ende, I make an other ge=mowe line, whiche is parallel to* N.F. *and* H.G, *and that ge=mowe line doth paſſe by the pricke* M, *and then haue I done. Now ſay I that* H.C.K.L, *is a likeiamme equall to the trian=gle appointed, whiche was* D, *and is made of a line aſſigned that is* A.B, *for* H.C, *is equall vnto* A.B, *and ſo is* K.L, *The profe of ỹ equalnes of this likeiam vnto the triāgle, depēdeth of the thirty and two Theoreme : as in the boke of Theoremes doth appear, where it is declared, that in al likeiammes, whē there are more then one made about one bias line, the fil ſqua=res of euery of them muſte needes be equall.*

E.   *The*

CONCLVSIONS

THE XVII. CONCLVSION.

To make a likeiamme equal to any right lined figure, and that on an angle appointed.

The readieſt waye to worke this concluſion, is to tourn that rightlined figure into triangles, and then for euery triangle to=gether an equal likeiamme, according vnto the eleuen cōcluſion, and then to ioine al thoſe likeiammes into one, if their ſi=des happen to be equal, which thing is euer certain, when al the triangles happē iuſtly betwene one pair of gemow lines. but and if they will not frame ſo, then after that you haue for the firſte triangle made his likeiamme, you ſhall take the lēgth of one of his ſides, and ſet that as a line aſſigned on whiche you ſhal make al the other likeiams, according to the twelft cō

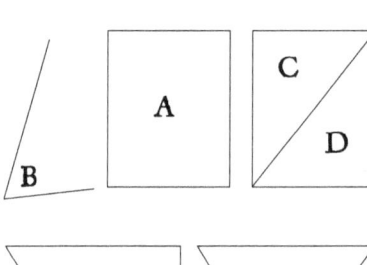

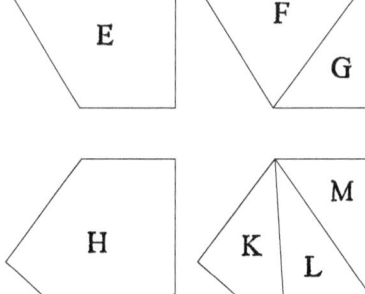

cluſion, and ſo ſhall you haue al your likeiammes with ij. ſides equal, and ij. like angles, ſo y̆ you mai eaſily ioyne thē into one figure.

Example.

If the right lined fi=gure be like vnto A, thē may it be turned into tri angles that wil ſtād be=twene ij. parallels anye ways, as you mai ſe by C and D, for ij. ſides of both the triāgls ar parallels. Alſo if the right lined fi gure be like vnto E, thē wil it be turned into triā gles, liyng betwene two parallels alſo, as y̆ other did before as in the exā=ple of F.G. But and if y̆ righte

GEOMETRICALL.

right lined figure be like vnto H, *and so turned into triāgles
as you se in* K.L.M, *wher it is parted into iij triāgles, thē wil
not all those triangles lye betwen one pair of parallels or ge=
mow lines, but must haue many, for euery triangle must haue
one paire of parallels feuerall, yet it maye happen that when
there bee three or fower triangles, ij. of theym maye hap=
pen to agre to one pair of parallels, whiche thinge I remit to
euery honest witte to ferche, for the manner of their draught
wil declare, how many paires of parallels they shall neede,
of which varietee bicaufe the examples ar infinite, I haue fet
forth thefe few, that by them you may coniecture duly of all
other like.*

*Further explicacion you shal not greatly neede, if you re=
membre what hath ben taught before, and then diligētly be=
hold how thefe fundry figures be turned into triāgles. In the
fyrst you fe I haue made* v. *triangles, and four paralleles. in
the feconde* vij. *triangles and foure paralleles. in the thirde
thre triāgles, and fiue parallels, in the iiij. you fe fiue triāgles
& four parallels. in the fift, iiij. triāgles and .iiij. parallels, & in y̆
fixt ther ar fiue triāgles & iiij. paralels. Howbeit a mā maye at
liberty alter them into diuers formes of triāgles & therefore I*

                                       E.ij.   *leue*

CONCLVSIONS

*leue it to the difcretion of the woorkmaifter, to do in al fuche
cafes as he fhal thinke beft, for by thefe examples (if they bee
well marked) may all other like conclufions be wrought.*

THE XVIII. CONCLVSION.

*To parte a line affigned after fuche a forte,
that the fquare that is made of the whole line
and one of his parts, fhal be equal to the fquar
that cometh of the other parte alone.*

*Firft deuide your lyne into ij. equal parts, and of the length
of one part make a perpendicular to light at one end of your
line affigned then adde a bias line, and make thereof a trian=
gle, this done if you take from this bias line the halfe lengthe
of your line appointed, which is the iufte length of your per=
pendicular, that part of the bias line whiche dothe remayne,
is the greater portion of the deuifion that you feke for, there=
fore if you cut your line according to the lengthe of it, then
will the fquare of that greater portion be equall to the fquare
that is made of the whole line and his leffer portion. And con
trary wife, the fquare of the whole line and his leffer parte,
wyll be equall to the fquare of the greater parte.*

Example.

A.B, *is the lyne affigned.* E. *is the
middle pricke of* A.B, B.C. *is the
plumb line or perpendicular, made
of the halfe of* A.B, *equall to* A.E,
other B.E, *the byas line is* C.A, *from
whiche I cut a peece, that is* C.D,
equall to C.B, *and accordyng to the
lengthe fo the peece that remaineth
(whiche is* D.A,) *I doo deuide the*

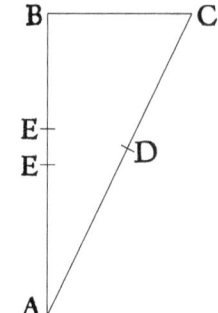

line A.B, *at whiche diuifion I fet* E. *Now fay I, that this line*
A,B, *(w was affigned vnto me) is fo diuided in this point* F, *y*
y fquare of y hole line A.B, *& of the one portiō (y is* F.B, *the*
leffer

GEOMETRICALL.

leſſer part) is equall to the ſquare of the other parte, whiche is F.A, *and is the greater part of the firſt line. The profe of this equalitie ſhall you learne by the .xl. Theoreme.*

THE XIX. CONCLVSION.

To make a ſquare quadrate equall to any right lined figure appointed.

    Firſt make a likeiamme equall to that right lined figure, with a right angle, accordyng to the .xi concluſion, then conſider the likeiamme, whether it haue all his ſides equall, or not : for yf they be all equall, then haue you doone your concluſion. but and if the ſides be not all equall, then ſhall you make one right line iuſte as long as two of thoſe vnequall ſides, that line ſhall you deuide in the middle, and on that pricke drawe half a cir= cle, then cutte from that diameter of the halfe circle a certayne portion equall to the one ſide of the likeiamme, and from that pointe of diuiſion ſhall you erecte a perpendicular, which ſhall touche the edge of the circle. And that perpendicular ſhall be the iuſte ſide of the ſquare quadrate, equall both to the lyke= iamme, and alſo to the right lined figure appointed, as the con= cluſion willed.   Example.

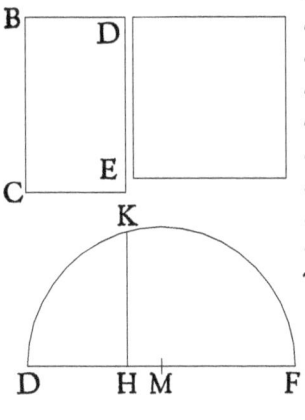

K *is the right lined figure appointed, and* B.C.D.E, *is the likeiãme, with right angles equall vnto* K, *but becauſe that this likeiamme is not a ſquare quadrate, I muſt turne it into ſuch one after this ſort, I ſhall make one right line, as long as .ij. vnequall ſides of the like= iãme, that line here is* F.G, *whiche is equall to* B.C, *and* C.E. *Then part I that line in the middle in the*
        E.iiij.   pricke

CONCLVSIONS

pricke M, *and on that pricke I make halfe a circle, accordyng to the length of the diameter* F.G. *Afterward I cut awaie a peece from* F.G, *equall to* C.E, *markyng that point with* H. *And on that pricke I erecte a perpendicular* H.K, *whiche is the iust side to the square quadrate that I seke for, therfore ac= cordyng to the doctrine of the .x. conclusion, of that lyne I doe make a square quadrate, and so haue I attained the practise of this conclusion.*

THE XX. CONCLVSION.

When any .ij. square quadrates are set forth, how you maie make one equall to them bothe.

First drawe a right line equall to the side of one of the qua= drates : and on the ende of it make a perpendicular, equall in length to the side of the other quadrate, then drawe a byas line betwene those .ij. other lines, makyng thereof a right angeled triangle. And that byas lyne wyll make a square quadrate, e= quall to the other .ij. quadrates appointed.

Example.

A.B. *and* C.D, *are the two square quadrates appointed, vnto which I must make one equall square quadrate. First therfore I doe make a righte line* E.F, *equall to one of the sides of the square quadrate* A.B. *And on the one end of it I make a plumbe line* E.G, *e= quall to the side of the other quadrate* D.C. *Then drawe I a byas line* G.F, *whiche be= yng made the side of a qua=*

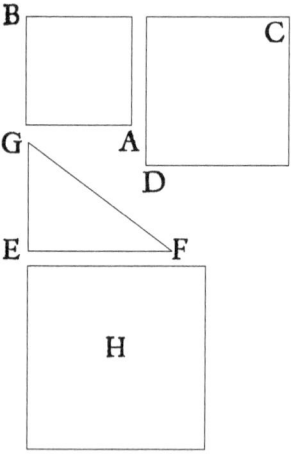

drate

GEOMETRICALL.

drate (accordyng to the tenth conclufion) will accomplifhe the worke of this practife : for the quadrate H. *is as muche iuft as the other two. I meane* A.B. *and* D.C.

THE XXI. CONCLVSION.

When any two quadrates be fet forth, howe to make a fquire about the one quadrate, whi= che fhall be equall to the other quadrate.

Determine with yourfelfe about whiche quadrate you wil make the fquire, and drawe one fide of that quadrate forth in lengte, accordyng to the meafure of the fide of the other qua= drate, whiche line you maie call the grounde line, and then haue you a right angle made on this line by an other fide of the fame quadrate : Therfore turne that into a right cornered tri= angle, accordyng to the worke in the lafte conclufion, by ma= kyng of a byas line, and that byas lyne will performe the worke of your defire. For if you take the length of that byas line with your compaffe, and then fet one foote of the compas in the far= theft angle of the firft quadrate (whiche is the one ende of the ground line) and extend the other foote on the fame line, accor= dyng to the meafure of the byas line, and of that line make a quadrate, enclofyng y̆ firft qua= drate, then will there appere the forme of a fquire about the firft quadrate, which fquire is equall to the fecond quadrate.

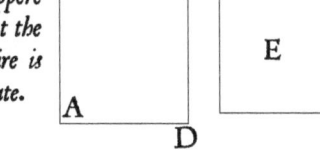

Example.

The firft fquare quadrate is A. B.C.D, *and the fecond is* E. *N̆ow would I make a fquire about the quadrate* A.B.C.D, *whiche fhall bee equall vnto the quadrate* E.

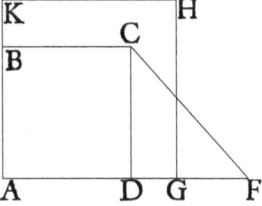

Therfore

CONCLVSIONS

Therfore firſt I draw the line A.D, *more in length, accordyng to the meaſure of the ſide of* E, *as you ſee, from* D. *vnto* F, *and ſo the hole line of bothe theſe ſeuerall ſides is* A.F, *thē make I a byas line from* C, *to* F, *whiche byas line is the meaſure of this woorke. wherefore I open my compas accordyng to the length of that byas line* C.F, *and ſet the one compas foote in* A, *and extend thother foote of the compas toward* F, *makyng this pricke* G, *from whiche I erect a plumbe line* G.H, *and ſo make out the ſquare quadrate* A.G.H.K, *whoſe ſides are e= quall eche of them to* A.G. *And this ſquare doth contain the firſt quadrate* A.B.C.D, *and alſo a ſquire* G.H.K, *whiche is equall to the ſecond quadrate* E, *for as the laſt concluſion de= clareth, the quadrate* A.G.H.K, *is equall to bothe the other quadrates propoſed, that is* A.B.C.D, *and* E. *Then muſte the ſquire* G.H.K, *needes be equall to* E, *conſideryng that all the reſt of that great quadrate is nothyng els but the quadrate ſelf,* A.B.C.D, *and ſo haue I thintent of this concluſion.*

THE XXI. CONCLVSION.

To find out the cētre of any circle aſſigned.

Draw a corde or ſtryng line croſſe the circle, then deuide in= to .ij. equall partes, both that corde, and alſo the bowe line, or arche line, that ſerueth to that corde, and from the prickes of thoſe diuiſions, if you drawe an other line croſſe the circle, it muſt nedes paſſe by the centre. Therfore deuide that line in the middle, and that middle pricke is the centre of the circle pro= poſed.

Example.

Let the circle be A.B.C.D, *whoſe centre I ſhall ſeke. Firſt therfore I draw a corde croſſe the circle, that is* A.C. *Then do I deuide that corde in the middle, in* E, *and likewaies alſo do I deuide his arche line* A.B.C, *in the middle, in the pointe* B. *Afterward I drawe a line from* B. *to* E, *and ſo croſſe the*
                                                                   circle

GEOMETRICALL.

circle, whiche line is B.D, *in which line is the centre that I ſeeke for. Therefore if I parte that line* B.D, *in the middle in to two equall portions, that middle pricke (whiche here is* F) *is the verye centre of the ſayde circle that I ſeke. This concluſion may otherwaies be wrought, as the moſte part of concluſions haue ſondry formes of practiſe, and that is, by ma= kinge thre prickes in the circū ference of the circle, at liberty where you wyll, and then fin= dinge the centre to thoſe thre prickes, which worke bicauſe it ſerueth for ſondry vſes, I thinke meet to make it a ſeuerall con cluſion by itſelfe.*

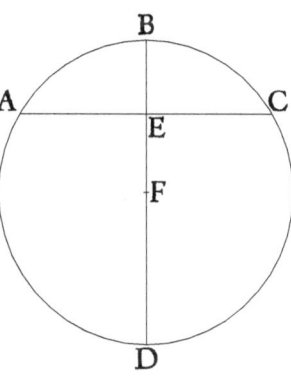

THE XXIII. CONCLVSION.

To find the commen centre belongyng to anye three prickes appointed, if they be not in an ex acte right line.

It is to be noted, that though euery ſmall arche of a greate circle do ſeeme to be a right lyne, yet in very dede it is not ſo, for euery part of the circumference of al circles is compaſſed, though in litle arches of great circles the eye cannot diſcerne the crokednes, yet reaſon doeth alwaies declare it, therfore iij. prickes in an exact right line can not bee brought into the circumference of a circle. But and if they be not in a right line how ſo euer they ſtande, thus ſhall you find their cōmon centre. Opē your compas ſo wide, that it be ſomewhat more then the

F.   *halfe*

CONCLVSIONS

halfe diſtance of two of thoſe prickes. Then ſette the one foote of the compas in the one pricke, and with the other foot draw an arche lyne toward the other pricke. Then againe putte the foot of your compas in the ſecond pricke, and with the other foot make an arche line, that may croſſe the firſte arch line in ij. places. Now as you haue done with thoſe two prickes, ſo do with the middle pricke, and the thirde that remayneth. Then draw ij. lines by the poyntes where thoſe arche lines do croſſe, and where thoſe two lines do meete, there is the cen=tre that you ſeeke for.   Example

The iij. prickes I haue ſet to be A.B, *and* C, *whiche I wold bring into the edg of one common circle, by finding a centre cōmen to them all, fyrſt therefore I open my cōpas, ſo that thei occupye more then y̆ halfe diſtance betwene ij. pricks (as are* A.B.) *and ſo ſet=tinge one foote in* A. *and extendinge the other to=ward* B, *I make the arche line* D.E. *Likewiſe ſettig̃ one foot in* B, *and turninge*

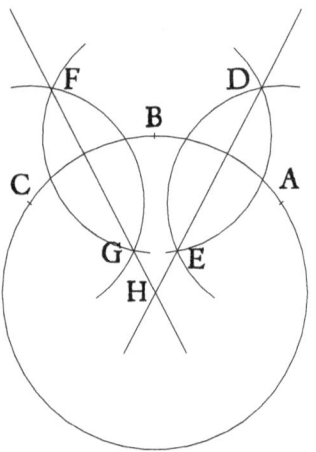

the other toward A, *I draw an other arche line that croſſeth the firſt in* D. *and* E. *Then from* D. *to* E, *I draw a right lyne* D.H. *After this I open my copaſſe to a new diſtance, and make ij. arche lines betwene* B. *and* C, *whiche croſſe one the other in* F. *and* G, *by whiche two pointes I draw an other line, that is* F.H. *And bycauſe that the lyne* D.H. *and the lyne* F.H. *doo meete in* H, *I ſaye that* H. *is the centre that ſerueth to thoſe iij. prickes. Now therfore, if you ſet one foot of your compas in* H, *and extend the other to any of the iij. pricks, you may draw a circle w̃ ſhal encloſe thoſe iij. pricks in the edg of his circūferēce, & thus haue you attained y̆ vſe of this cōcluſiō*
                                                                                          The

GEOMETRICALL.

THE XXIIII. CONCLVSION.

To drawe a touche line vnto a circle, from any pointe aſſigned.

Here muſt you vnderſtand that the pricke muſt be without the circle, els the concluſion is not poſſible. But the pricke or point beyng without the circle, thus ſhall you procede : Open your compas, ſo that the one foote of it maie be ſet in the centre of the circle, and the other foote on the pricke appointed and ſo draw an other circle of that largeneſſe about the ſame cen= tre : and it ſhall gouerne you certainly in makyng the ſaid tou= che line. For if you draw a line frō the pricke appointed vn= to the centre of the circle, and marke the place where it doeth croſſe the leſſer circle, and from that pointe erect a plumbe line tht ſhall touche the edge of the vtter circle, and marke alſo the place where that plumbe line croſſeth that vtter cir= cle, and from that place drawe an other line to the centre, ta= kyng heede where it croſſeth the leſſer circle, if you drawe a plumbe line from that pricke vnto the edge of the greatter circle, that line I ſay is a touche line, drawen from the point aſſigned, according to the meaning of this concluſion.

Example.

Let the circle be called B.C. D, and his cētre E, and ẙ prick aſſigned A, opē your cōpas now of ſuch widenes, ẙ the one foote may be ſet in E, w̃ is ẙ cētre of ẙ circle, & ẙ other in A, w̃ is ẙ pointe aſſigned, & ſo make an o= ther greter circle (as here is A.F. G) thē draw a line from A. vnto E, and wher that line doth croſſ ẙ inner circle (w̃ heere is in the pricke B.) there erect a plūb line vnto the line. A.E. and let that

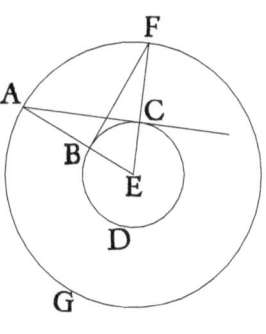

plumb line touch the vtter circle, as it doth here in the point F, ſo ſhall B.F. bee that plumbe lyne. Then from F. vnto E.

F.ij.   draw

CONCLVSIONS

drawe an other line whiche ſhal be F.E, and it will cutte the
inner circle, as it doth here in the point C, from which pointe
C. if you erect a plumb line vnto A, then is that line A.C, the
touche line, whiche you ſhoulde finde. Notwithſtandinge
that this is a certaine waye to fynde any touche line, and a
demonſtrable forme, yet more eaſyly by many folde may you
fynde and make any ſuche line with a true ruler, layinge the
edge of the ruler to the edge of the circle and to the pricke,
and ſo drawing a right line, as this example ſheweth, where
the circle is E, the pricke aſſiː
gned is A. and the ruler C.D.
by which the touchline is dra
wen, and that is A,B, and as
this way is light to doo, ſo is it
certaine inoughe for any kinde
of workinge.

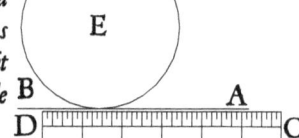

THE XXV. CONCLVSION.

When you haue any peece of the circumference
of a circle aſſigned, howe you may make oute
the whole circle agreynge thereunto.

Firſt ſeeke out the centre of that arche, according to the doc
trine of the ſeuententh concluſion, and then ſetting one foote
of your compas in the centre, and extending the other foot vn
to the edge of the arche or peece of the circumference, it is eaː
ſy to drawe the whole circle.

Example.

A peece of an olde piller was found, like in forme to thys
figure A.D.B. Now to knowe howe muche the cōpaſſe of
the hole piller was, ſeing by this parte it appereth that it was
round, thus ſhal you do. Make in A table the like draught of ẏ
circūference by the ſelf patrō, vſing it as it wer a croked ruler.
*   Then*

GEOMETRICALL.

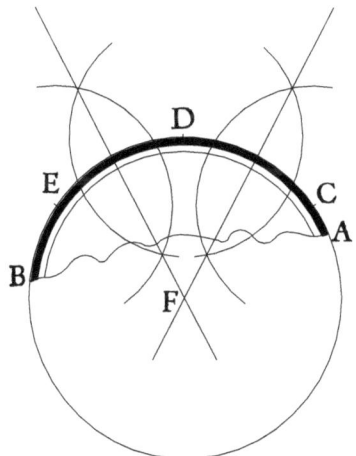

Then make .iij. prickes in that arche line, as I haue made, C.D. and E. And then finde out the common centre to them all, as the .xvij. conclusion teacheth. And that cētre is here F, nowe settyng one foote of your compas in F, and the other in C.D, other in E, and so makyng a com= passe, you haue youre whole intent.

THE XXVI. CONCLVSION.

To finde the centre to any arche of a circle.

If so be it that you desire to find the centre by any other way then by those .iij. prickes, consideryng that sometimes you can not haue so muche space in the thyng where the arche is dra= wen, as should serue to make those .iiij. bowe lines, then shall you do thus : Parte that arche line into two partes, equall o= ther vnequall, it maketh no force, and vnto ech portion draw a corde, other a stringline. And then accordyng as you dyd in one arche in the .xvi. conclusion, so doe in bothe those arches here, that is to saie, deuide the arche in the middle, and also the corde, and drawe then a line by those two deuisions, so then are you sure that that line goeth by the centre. Afterward do likewaies with the other arche and his corde, and where those .ij. lines do crosse, there is the centre, that you seke for.

Example.

The arche of the circle is A.B.C, vnto whiche I must seke
F.iij.   a cen=

CONCLVSIONS

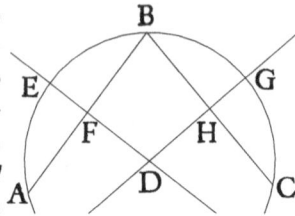

a centre, therfore firſte I do
deuide it into .ij. partes, the
one of them is A.B, *and the*
other is B.C. *Then doe I cut*
euery arche in the middle, ſo
is E. *the middle of* A.B, *and*
G. *is the middle of* B.C. *Like-*
waies, I take the middle of their cordes, whiche I mark with
F. *and* H, *ſettyng* F. *by* E, *and* H. *by* G. *Then drawe I a*
line from E. *to* F, *and from* G. *to* H, *and they do croſſe in* D,
wherefore ſaie I, that D. *is the centre, that I ſeke for.*

THE XXVII. CONCLVSION.

To drawe a circle within a triangle ap=
poinɛted.

For this concluſion and all other lyke, you muſte vnder=
ſtande, that when one figure is named to be within an other,
that is not other waies to be vnderſtande, but that eyther eue=
ry ſyde of the inner figure dooeth touche euerie corner of the
other, other els euery corner of the one dooeth touche euerie
ſide of the other. So I call that triangle drawen in a circle,
whoſe corners do touche the circumference of the circle. And
that circle is contained in a triangle, whoſe circumference do=
eth touche iuſtely euery ſide of the triangle, and yet dooeth
not croſſe ouer any ſide of it. And ſo that quadrate is called
properly to be drawen in a circle, when all his fower angles
dooeth touche the edge of the circle. And that circle is drawen
in a quadrate, whoſe circumference dooeth touche euery ſide of
the quadrate, and lykewaies of other figures.

Ex=

GEOMETRICALL.

Examples are thefe. A. B. C. D. E. F.

A. is a circle
in a triangle.

C. a quadrate
in a circle.

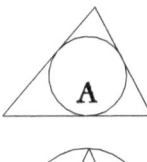

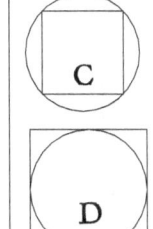

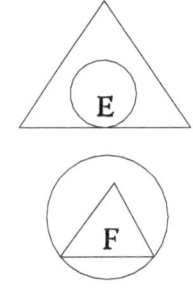

B. a triangle
in a circle.

D. a circle in
a quadrate.

In thefe .ij. laſt figures E. and F, the circle is not named to be drawen in a triangle, becauſe it doth not touche the ſides of the triangle, neither is the triangle coūted to be drawen in the circle, becauſe one of his corners doth not touche the circum= ference of the circle, yet (as you fee) the circle is within the tri= angle, and the triangle within the circle, but nother of them is properly named to be in the other. Now to come to the con= cluſion. If the triangle haue all .iij. ſides lyke, then ſhall you take the middle of euery ſide, and from the contrary corner drawe a right line vnto that poynte, and where thoſe lines to croſſe one another, there is the centre. Then ſet one foote of the compas in the centre, and ſtretche out the other to the mid= dle pricke of any of the ſides, and ſo drawe a compas, whiche ſhall touche euery ſide of the triangle, but ſhall not paſſe with out any of them.   Example.

The triangle is A.B.C, whoſe ſides I do part into .ij. equall partes, eche by it ſelfe in theſe pointes D.E.F, puttyng F. be= twene A.B, and D. betwene B.C, and E. betwene A.C. Then draw I a line from C. to F, and an other from A. to D, and the third from B. to E.

                                      And

CONCLVSIONS

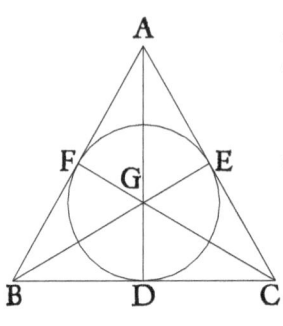

And where all thofe lines do mete (that is to faie in G,) I fet the one foote of my com= paffe, becaufe it is the com= mon centre, and fo drawe a circle accordyng to the di= ftaunce of any of the fides of the triangle. And then find I that circle to agree iuftely to all the fides of the triangle, fo that the circle is iuftely made in the triangle, as the conclu= fion did purporte. And this is euer true, when the triangle hath all thre fides equall, other at the leaft .ij. fides lyke long. But in the other kindes of triangles you muft deuide euery an= gle in the middle, as the third conclufion teacheth you. And fo drawe lines frō eche angle to their middle pricke. And where thofe lines do croffe, there is the common centre, from which you fhall draw a perpendicular to one of the fides. Then fette one foote of the compas in that centre, and ftretche the other foote accordyng to the lēgth of the perpendicular, and fo draw your circle.

Example.

The triangle is A.B.C, *whofe corners I haue diui= ded in the middle with* D. E.F, *and haue drawen the li= nes of diuifion* A.D.B.E. *and* C.F, *whiche croffe in* G, *therfore fhall* G. *be the common centre. Then make I one perpēdicular from* G. *vnto the fide* A.C, *and that*

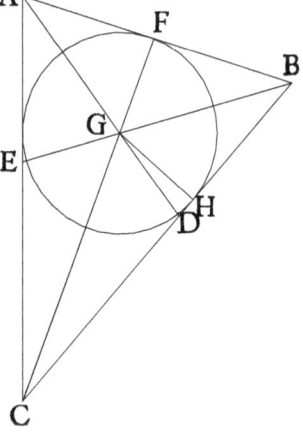

is

GEOMETRICALL.

is G.H. *Nowe ſette I one fote of the compas in* G, *and extend the other foote vnto* H. *and ſo drawe a compas, whiche wyll iuſtly anſwere to that triãgle according to the meaning of the concluſion.*

THE XXVIII. CONCLVSION.

To drawe a circle about any triãgle aſſigned.

Fyrſte deuide two ſides of the triangle equally in half, and from thoſe ij. prickes erect two perpendiculars, which muſte needes meet in croſſe, and that point of their meting is the cen tre of the circle that muſt be drawen, therefore ſette one foote of the compaſſe in that pointe, and extend the other foote to one corner of the triangle, and ſo make a circle, and it ſhall touche all iiij. corners of the triangle.

Example.

A.B.C. *is the triangle, whoſe two ſides* A.C. *and* B,C. *are diuided into two e= quall partes in* D. *and* E, *ſettyng* D. *be= twene* B. *and* C, *and* E. *betwene* A. *and* C. *And from eche of thoſe two pointes is ther erected a perpendicular (as you ſe* D.F, *and* E.F.) *which mete, and croſſe in* F, *and ſtretche forth the other foot of any corner of the triangle, and ſo make a circle, that circle ſhal touch euery cor= ner of the triangle, and ſhal encloſe the whole triangle, accor dinge, as the concluſion willeth.*

An other waye to do the ſame.

And yet an other waye may you doo it, accordinge as you learned in the ſeuententh concluſion, for if you call the three
     corner

CONCLVSIOSN

corners of the triangle iij. prickes, and then (as you learned there) yf you seeke out the centre to thofe three prickes, and fo make it a circle to inclofe thofe thre prickes in his circum= ference, you fhall perceaue that the fame circle fhall iuftelye include the triangle propofed.

Example.

A.B.C. is the triangle, whofe iij. cor= ners I count to be iij. pointes. Then (as the feuentene conclufion doth teache) I feeke a common centre, on which I may make a circle, that fhall enclofe thofe iij prickes that centre, as you fe is D, for in D. doth the right lines, that paffe by the angles of the arche lines, meete and croffe. And on that centre as you fe, haue I made a circle, which doth inclofe the iij. angles of the triãgle, and confequent lye the triangle itfelfe, as the conclufion dydde intende.

THE XXIX. CONCLVSION.

To make a triangle in a circle appoynted whofe corners fhalbe equall to the corners of any triangle affigned.

When I will draw a triangle in a circle appointed, fo that the corners of that triangle fhall be equall to the corners of a= ny triangle affigned, then muft I firft draw a tuche lyne vn to that circle, as the twenty conclufion doth teach, and in the very poynte of the touche mufte I make an angle, equall to one angle of the triangle, and that inwarde toward the cir= cle : likewife in the fame pricke mufte I make an other angle w the other halfe of the touche line, equall to an other corner of the triangle appointed, and then betwen thofe two corners

wyl

GEOMETRICALL.

will there refulte a third angle, equall to the third corner of that triangle. Nowe where thofe two lines that entre into the circle, doo touche the circumference (befide the touche line) there fet I two prickes, and betwene them I drawe a thyrde line. And fo haue I made a triangle in a circle appointed, whofe corners bee equall to the corners of the triangle affi= gned.

Example.

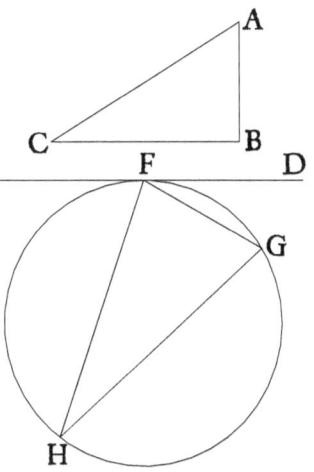

A.B.C, *is the triangle appointed, and* **F.G.H.** *is the circle, in which I mufte make an other triangle, with lyke angles to the angles of* **A.B.C.** *the triangle ap pointed. Therefore fyrft I make the touch lyne* **D.F.E.** *And then make I an angle in* **F,** *equall to* **A,** *whiche is one of the angles of the triangle. And the lyne that maketh that angle with the touche line, is* **F.H,** *whiche I drawe in lengthe vntill it touche the edge of the circle. Then againe in the fame point* **F,** *I make an other corner equall to the angle* **C.** *and the line that maketh that corner with the touche line, is* **F.G.** *whiche alfo I drawe foorthe vntill it touche the edge of the circle. And then haue I made three angles v= pon that one touch line, and in ẙ one point* **F,** *and thofe iij. an= gles be equal to the iij. angles of the triangle affigned, whiche thinge doth plainely appeare, in fo muche as they bee equall*

G.ij   to

CONCLVSIONS

to ij. right angles, as you may geſſe by the ſixt theoreme. And
the thre angles of euerye triangle are equall alſo to ij. righte
angles, as the two and twenty theoreme dothe ſhow, ſo that
bicauſe they be equall to one thirde thinge, they muſt needes
be equal togither, as the cōmon ſentence ſaith. Thē do I draw
a line from G. to H, and that line maketh a triangle F.G.H.
whoſe angles be equall to the angles of the triangle appoin=
ted. And this triangle is drawen in a circle, as the concluſion
didde wyll. The proofe of this concluſion doth appeare in the
ſeuenty and iiij. Theoreme.

THE XXX. CONCLVSION.

To make a triangle about a circle aſſigned
whiche ſhall haue corners, equall to the cor=
ners of any triangle appointed.

Firſt draw forth in length the one ſide of the triangle aſſigned
ſo that therby you may haue ij. vtter angles, vnto which two
vtter angles you ſhall make ij. other equall on the centre of
the circle propoſed, drawing thre halfe diameters frome the
circumference, whiche ſhal encloſe thoſe ij. angles, thē draw
iij. touche lines which ſhall make ij. right angles, eche of them
with one of thoſe ſemidiameters. Thoſe iij. lines will make
a triangle equally cornered to the triangle aſſigned, and that tri
angle is drawē about a circle apointed, as the cōcluſiō did wil.

Example.

A.B.C, is the triangle aſſigned, and G.H.K, is the circle ap
pointed, about which I muſte make a triangle hauing equall
angles to the angles of that triangle A.B.C. Fyrſt therefore I
draw A.C. (which is one of the ſides of the triangle) in length
that there may appeare two vtter angles in that triangle, as
you ſe B.A.D, and B.C. E.

Then

GEOMETRICALL.

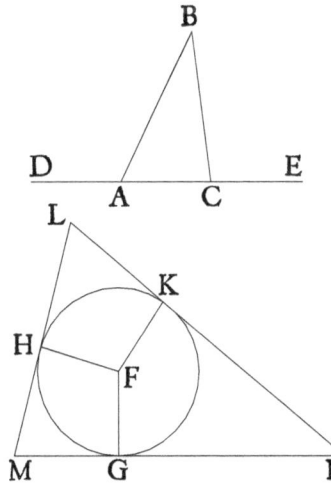

Then drawe I in the circle appointed a ſe=midiameter, which is here H.F, for F. is the cētre of the circle G. H.K. Then make I on that centre an angle e=quall to the vtter an=gle B.A.D, and that angle is H.F.K. Like waies on the ſame cē=tre by drawyng an o=ther ſemidiameter, I make an other angle H.F.G, equall to the ſecond vtter angle of the triangle, whiche is B.C.E. *And thus haue I made .iij. ſe=midiameters in the circle appointed. Then at the ende of eche ſemidiameter, I draw a touche line, whiche ſhall make righte angles with the ſemidiameter. And thoſe .iij. touch lines mete, and you ſee, and make the triangle L.M.N, whiche is the tri=angle that I ſhould make, for it is drawen about a circle aſſig=ned, and hath corners equall to the corners of the triangle ap=pointed, for the corner M. is equall to C. Likewaies L. to A, and N. to B, whiche thyng you ſhall better perceiue by the vi. Theoreme, as I will declare in the booke of proofes.*

THE XXXI. CONCLVSION.

To make a portion of a circle on any right line aſſigned, whiche ſhall conteine an angle e=quall to a right lined angle appointed.

The angle appointed, maie be a ſharpe angle, a right angle, o=ther a blunte angle, ſo that the worke muſt be diuerſely han=

CONCLVSIONS

deled *according to the diuerſities of the angles, but conſide=*
ringe the hardenes of thoſe ſeuerall woorkes, I wyll omitte
them for a more meter time, and at this tyme wyll ſhewe you
one light waye which ſerueth for all kindes of angles, and that
is this. When the line is propoſed, and the angle aſſigned, you
ſhall ioyne that line propoſed ſo to the other twoo lines con=
tayninge the angle aſſigned, that you ſhall make a triangle
of theym, for the eaſy dooinge whereof, you may enlarge
or ſhorten as you ſee cauſe, nye of the two lynes contayninge
the angle appointed. And when you haue made a triangle
of thoſe iij. lines, then accordinge to the doctrine of the ſeuē
and twēty cōcluſiō, make a circle about that triangle. And ſo
haue you wroughte the requeſt of this concluſion. whyche
yet you maye woorke by the twenty and eight concluſion alſo,
ſo that of your line appointed, you make one ſide of the triā=
gle be equal to
ỹ āgle aſſigned
as youre ſelfe
mai eaſily geſſe

Example.

Firſt for exam=
ple of a ſharpe
āgle let A. ſtād
& B.C. ſhal be ỹ
line   aſſigned.
Thē do I make
a triangle, by
adding B.C, as
a thirde ſide to
thoſe other ij.
which doo in=
clude the āgle
aſſigned,   and
that triāgle is D
E.F, ſo ỹ E.F.
is

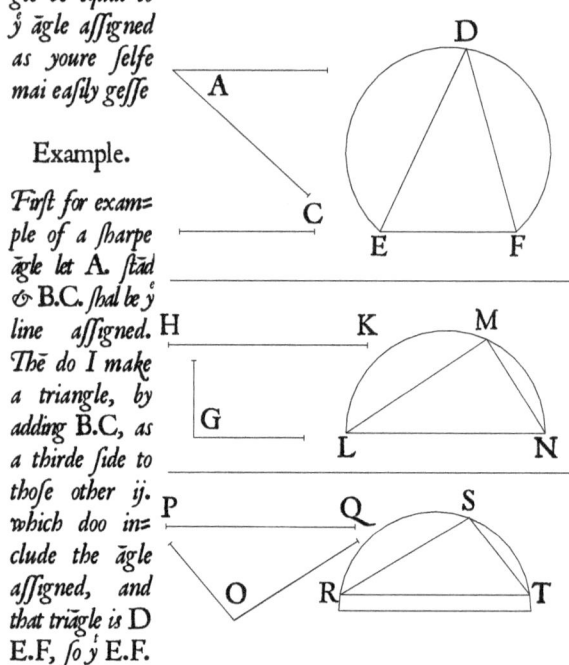

GEOMETRICALL.

is the line appointed, and D. *is the angle assigned. Then doe I
drawe a portion of a circle about that triangle, from the one
ende of that line assigned vnto the other, that is to saie, from*
E. *a long by* D. *vnto* F, *whiche portion is euermore greatter
then the halfe of the circle, by reason that the angle is a sharpe
angle. But if the angle be right (as in the second exaumple you
see it) then shall the portion of the circle that containeth that
angle, euer more be the iuste halfe of a circle. And when the
angle is a blunte angle, as the thirde exaumple dooeth pro=
pounde, then shall the portion of the circle euermore be lesse
then the halfe circle. So in the seconde example,* G. *is the right
angle assigned, and* H.K. *is the lyne appointed, and* L.M.N.
*the portion of the circle aunsweryng thereto. In the third ex=
aumple,* O. *is the blunte corner assigned,* P.Q. *is the line, and*
R.S.T. *is the portion of the circle, that containeth that blūt
corner, and is drawen on* R.T. *the lyne appointed.*

THE XXXII. CONCLVSION.

*To cutte of from any circle appoineed, a
portion containyng an angle equall to a right
lyned angle assigned.*

*When the angle and the circle are assigned, first draw a touch
line vnto that circle, and then drawe an other line from the
pricke of the touchyng to one side of the circle, so that thereby
those two lynes do make an angle equall to the angle assigned.
Then saie I that the portion of the circle of the contrarie side
to the angle drawen, is the parte that you seke for.*

Example.

A. *is the angle appointed, and* D.E.F. *is the circle assigned,
frō which I must cut away a portiō that doth contain an angle*
                                                          equall

CONCLVSIONS

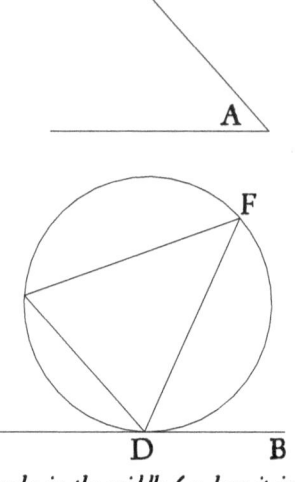

equall to this angle A.
Therfore firft I do draw
a touche line to the cir=
cle affigned, and that
touchline is B.C, the ve=
ry pricke of the touche is
D, from whiche D. I
drawe a lyne D.E, fo
that the angle made of
thofe two lines be equall
to the angle appointed.
Then fay I, that the arch
of the circle D.F.E, is
the arche that I feke af=
ter. For if I doo deuide
that arche in the middle (as here it is done in F.) and fo draw
thence two lines, one to A, and the other to E, then will the
angle F, be equall to the angle affigned.

THE XXXIII. CONCLVSION.

To make a fquare quadrate in a circle affigned.

Draw .ij. diameters in the circle, fo that they runne a croffe,
and that they make iiij. right angles. Then drawe .iiij. lines,
that may ioyne the .iiij. endes of thofe diameters, one to an o=
ther, and then haue you made a fquare quadrate in the circle
appointed.   Example.

A.B.C.D. *is the circle affig=*
ned, and A.C. *and* B.D. *are the*
two diameters whiche croffe in
the centre E, *and make .iiij. right*
corners. Then do I make fowre
other lines, that is A.B, B.C,
C.D, *and* D.A, *which do ioyne*
together the fowre endes of the
ij. diameters. And fo is the fquare

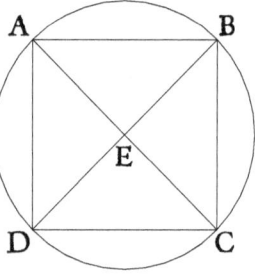

qua=

GEOMETRICALL.

quadrate made in the circle affigned, as the conclufion wil= leth.

THE XXXIIII. CONCLVSION.

To make a fquare quadrate aboute annye circle affigned.

Drawe two diameters in croffe waies, fo that they make foure righte angles in the centre. Then with your compaffe take the length of the halfe diameter, and fet one foote of the compas in eche end of thofe diameters, drawing twoo arche lines at euery pitchinge of the compas, fo fhall you haue viij. arche lines. Then yf you marke the prickes wherin thofe arch lines do croffe, and draw betwene thofe iiij. prickes iiij. right lines, then haue you made the fquare quadrate accordinge to the requeft of the conclufion.

Example.

A.B.C. *is the circle affigned* in which firft I draw two diameters, in croffe waies, making iiij. righte angles, and thofe ij. diameters are A.C. and B.D. Then fette I my compaffe (whiche is o= pened according to the fe= midiameter of the faid cir= cle) fixing one foote in the end of euery femidiameter, and drawe with the other foote twoo arche lines, one on euery fide. As firfte, when I fette the one foote in A,

H   then

CONCLVSIONS

then with the other foote I doo make twoo arche lines, one
in E, and an other in F. Then sette I the one foote of the com=
passe in B, and drawe twoo arche lines F. and G. Likewise
settinge the compasse foote in C, I drawe twoo other arche
lines, G. and H, and on D. I make twoo other, H. and E. Then
frome the crossinges of those eighte arche lines I drawe iiij.
straighte lynes, that is to saye, E.F, and F.G. also G.H, and
H.E, whiche iiij. straight lines do make the square quadrate
that I should draw about the circle assigned.

THE XXXV. CONCLVSION.

To drawe a circle in any square quadratae
appointed.

Fyrste deuide euery side of the quadrate into twoo equall
partes, and so drawe two lynes betwene eche two contrary
poinctes, and where those twoo lines doo crosse, there is the
centre of the circle. Then sette the one foote of the compasse
in that point, and stretch forth the other foot, according to the
length of halfe one of those lines, and so make a compas in the
square quadrate assigned.

Example.

A.B.C.D. is the quadrate appoin=
ted, in whiche I muste make a circle.
Therfore first I do deuide euery side
in ij. equal partes, and draw ij lines,
acrosse, betwene eche ij. cōtrary pric=
kes, as you se E.G, and F.H, whiche
mete in K, and therfore shil K, be the
centre of the circle. Then do I set one
foote of the compas in K. and opē the
other as wide as K.E, and so draw a
circle, whiche is made ancordinge to
the conclusion.

The

GEOMETRICALL.

THE XXXVI. CONCLVSION.

To draw a circle about a square quadrate.

Draw ij. lines betwene the iiij. corners of the quadrate, and where they mete in croſſe, ther is the centre of the circle that you ſeeke for. Thē ſet one foot of the compas in that centre, and extend the other foote vnto one corner of the quadrate, and ſo may you draw a circle which ſhall iuſtely incloſe the qua= drate propoſed.

Example.

A.B.C.D. *is the ſquare quadrate pro= poſed, about which I muſt make a cir= cle. Therfore do I draw ij. lines croſſe the ſquare quadrate from angle to an gle, as you ſe* A.C. & B.D. *And where they ij. do croſſe (that is to ſay in* E.) *there ſet I the one foote of the compas as in the centre, and the other foote I do extend vnto one angle of the qua= drate, as for exāple to* A, *and ſo make a compas, whiche doth iuſtly incloſe the quadrate, according to the minde of the concluſion.*

THE XXXVII. CONCLVSION.

To make a twileke triangle, whiche ſhall haue euery of the ij. angles that lye about the ground line, double to the other corner.

Fyrſte make a circle, and deuide the circumference of it into fyue equall partes. And thenne drawe frome one pricke (which you will) two lines to ij. other prickes, that is to ſay to the iij. and iiij. pricke, counting that for the firſt, wherhence you drewe both thoſe lines, Then drawe the thyrde lyne to make a triangle with thoſe other twoo, and you haue doone according to the concluſion, and haue made a twelike triāgle,

H.ij   *whoſe*

CONCLVSIONS

whoſe ij. corners about the grounde line, are eche of theym double to the other corner.

Example.

A.B.C. *is the circle, whiche I haue deuided into fiue equal por tions. And from one of the pric= kes (which is* A,*) I haue drawē ij. lines,* **A.B.** *and* **B.C,** *whiche are drawen to the third and iiij. prickes. Then draw I the third line* **C.B,** *which is the grounde line, and maketh the triangle, that I would haue, for the āgle* **C.** *is double to the angle* **A,** *and ſo is the angle* **B.** *alſo.*

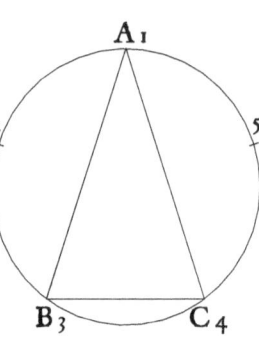

THE XXXVII. CONCLVSION.

To make a cinkangle of equall ſides, and equall corners in any circle appointed.

Deuide the circle appointed into fiue equall partes, as you didde in the laſte concluſion, and drawe ij. lines from euery pricke to the other ij. that are nexte vnto it. And ſo ſhall you make a cinkangle after the meanynge of the concluſion.

Example.

Yow ſe here this circle **A.B.C.D.E.** *deuided into fiue e= quall portions. And from eche pricke ij. lines drawen to the other ij. nexte prickes, ſo from* **A.** *are drawen ij. lines, one to* **B,** *and the other to* **E,** *and ſo from* **C.** *one to* **B.** *and an other*

to

GEOMETRICALL.

to D, *and likewise of the reste. So that you haue not on ly learned hereby how to make a finkangle in anye circle, but also how you fhal make a like figure fpedely, whanne and where you will, onlye drawinge the cir= cle for the intente, readylye to make the other figure (I meane the cinkangle) thereby.*

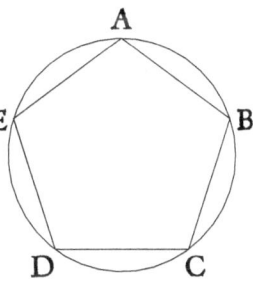

THE XXXIX. CONCLVSION.

How to make a cinkangle of equall fides and equall angles about any circle appointed.

Deuide firfte the circle as you did in the lafte conclufion in= to fiue equall portions, and draw fiue femidiameters in the circle. Then make fiue touche lines, in fuche forte that euery touche line make two right angles with one of the femidia= meters. And thofe fiue touche lines will make a cinkangle of equall fides and equall angles.

Example.

A.B.C.D.E. *is the circle appointed, which is deuided into fiue equal partes. And vnto euery prycke is drawē a femidiameter, as you fee. Then doo I make a touche line in the pricke* B, *whiche is* F.G, *makinge ij. right an=*

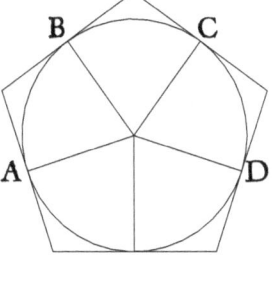

H.iij.   *gles*

CONCLVSIONS

gles with the femidiameter B, and lyke waies on C. is made
G.H, on D. ftandeth H.K, and on E, is fet K.L, fo that of
thofe .v. touche lynes are made the .v. fides of a cinkeangle,
accordyng to the conclufion.

An other waie.

Another waie alfo maie you drawe a cinkeangle aboute a
circle, drawyng firft a cinkeangle in the circle (whiche is an
eafie thyng to doe, by the doctrine of the .xxxvij. conclufion)
and then drawyng .v. touche lines whiche fhall be iufte paral=
leles to the .v. fides of the cinkeangle in the circle, forfeeyng
that one of them do not croffe ouerthwarte an other and then
haue you done. The exaumple of this (becaufe it is eafie) I
leaue to your owne exercife.

THE XL. CONCLVSION.

To make a circle in any appointed cinke=
angle of equall fides and equall corners.

Drawe a plumbe line from any one corner of the cinkeangle,
vnto the middle of the fide that lieth iufte againft that angle.
And do likewaies in drawyng an other line from fome other
corner, to the middle of the fide that lieth againft that corner
alfo. And thofe two lines wyll meete in croffe in the pricke
of their croffyng fhall you iudge the centre of the circle to be.
Therfore fet one foote of the compas in that pricke, and ex=
tend the other to the ende of the line that toucheth the middle
of one fide, whiche you lifte, and fo drawe a circle. And it
fhall be iuftly made in the cinkeangle, accordyng to the conclu=
fion.

Example.

The cinkeangle affigned is A.B.C.D.E, in whiche I mufte
make

GEOMETRICALL.

*make a circle, wherfore I draw a right line from the one an=
gle (as from* B,) *to the middle of the contrary side (whiche is*
E.D,) *and that middle pricke is* F. *Then lykewaies from an
other corner (as from* E) *I drawe a right line to the middle of
the side that lieth againſt it (whiche is* B.C.) *and that pricke is*

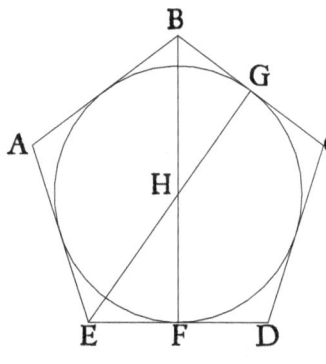

G. *Nowe becauſe that
theſe two lines do croſſe
in* H, *I ſaie that* H. *is the
centre of the circle, whi=
che I would make. Ther=
fore I ſet one foote of the
compaſſe in* H, *and extend
the other foote vnto* G, *or*
F. *(whiche are the endes
of the lynes that lighte in
the middle of the ſide of
that cinkeangle) and ſo
make I a circle in the cinkangle, right as the cōcluſion meaneth.*

THE XLI. CONCLVSION.

*To make a circle about any aſſigned cinke=
angle of equall ſides, and equall corners.*

*Drawe .ij. lines within the cinkeangle, from .ij. corners to the
middle on the .ij. contrary ſides (as the laſt concluſion teacheth)
and the pointe of their croſſyng ſhall be the centre of the cir=
cle that I ſeke for. Then ſette I one foote of the compas in that
centre, and the other foote I extend to one of the angles of the
cinkangle, and ſo draw I a circle about the cinkangle aſſigned.*

Example.

A.B.C.D.E, *is the cinkangle aſſigned, about which I would
make a circle, Therfore I drawe firſte of all two lynes (as you
ſee) one frō* E. *to* G, *and the other frō* C. *to* F, *and becauſe thei do*

                                                                                           meete

CONCLVSIONS

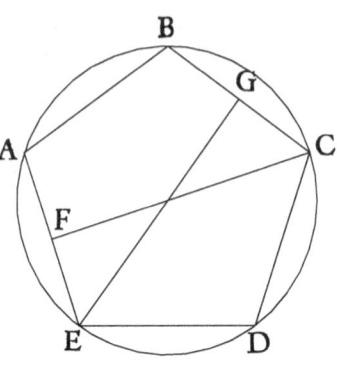

meete *in* H, *I ſaye that* H. *is the centre of the circle that I woulde haue, wherfore I ſette one foote* A *of the compaſſe in* H. *and extende the other to one corner (whiche happeneth fyrſte, for all are like diſtaunte from* H.*) and ſo make I a circle aboute the cinkeangle aſſigned.*

An other waye alſo.

An other waye maye I do it, thus preſuppoſing any three corners of the cinkangle to be three prickes appointed, vnto whiche I ſhoulde finde the centre, and then drawinge a circle touchinge them all thre, according to the doctrine of the ſeuentene, one and twenty, and two and twenty concluſions. And when I haue founde the centre, then doo I drawe the circle as the ſame concluſions do teache, and this forty concluſion alſo.

THE XLII. CONCLVSION.

To make a ſiſeangle of equall ſides, and equall angles, in any circle aſſigned.

If the centre of the circle be not knowen, then ſeeke oute the centre according to the doctrine of the ſixetenth concluſion. And with your compas take the quantitee of the ſemidiameter iuſtly. And then ſette one foote in one pricke of the circum=

GEOMETRICALL.

circūference of the circle, and with the other make a marke
in the circumference alſo towarde both ſides. Then ſette one
foote of the compas ſtedily in eche of thoſe newe prickes, and
point out two other prickes. And if you haue done well, you
ſhal perceaue that there will be but euen ſixe ſuch diuiſions in
the circumference. Whereby it dothe well appeare, that
the ſide of anye fiſeangle made in a circle, is equalle
to the ſemidiameter of the ſame circle.

Example.

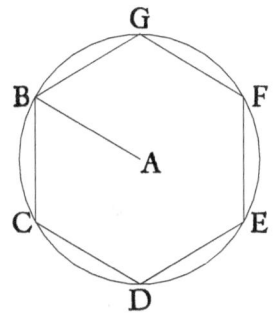

The circle is B.C.D.E.F.G,
whoſe centre I finde to bee A.
*Therefore I ſette one foote of the
compas in* A, *and do exted the o=
ther foote to* B, *thereby takinge
the ſemidiameter. Then ſette I
one foote of the compas vnremo=
ued in* B, *and marke with the o=
ther foote on eche ſide* C. *and* G.
Then from C. *I marke* D, *and frō*
D, E: *from* E. *marke I* F. *And then haue I but one ſpace iuſte
vnto* G. *and ſo haue I made a iuſte fiſeangle of equall ſides
and equall angles, in a circle appointed.*

THE XLIII. CONCLVSION.

*To make a fiſeangle of equall ſides, and e=
quall angles about any circle aſſigned.*

THE XLIIII. CONCLVSION.

*To make a circle in any fiſeangle appoin=
ted, of equall ſides and equal angles.*
L.i.   The

CONCLVSIONS

THE XLV. CONCLVSION.

To make a circle about any fifeangle limi=
ted of equall fides and equall angles,

Bicaufe you maye eafily conieƈure the makinge of thefe
figures by that that is faide before of cinkangles, only confide=
ringe that there is a difference in the numbre of the fides, I
thought befte to leue thefe vnto your owne deuice that you
fhould ftudy in fome thinges to exercife your witte withall
and that you mighte haue the better occafion to per=
ceaue what difference there is betwene eche twoo of thofe
conclufions. For thoughe it feeme one thing to make a fifean=
gle in a circle, and to make a circle about a fifeangle, yet fhill
you perceaue, that it is not one thinge, nother are thofe twoo
conclufions wrought one way. Likewaife fhall you thinke
of thofe other two conclufions. To make a fifeangle about a
circle, and to make a circle in a fifeangle, thoughe the figures
be one in fafhion, when they are made, yet are they not one
in working, as you may well perceaue by the xxxvij. xxxviiij
xxxix. and xl. conclufions, in whiche the fame workes are
taught, touching a circle and a cinkangle, yet this muche wyll
I faye, for your helpe in workyng, that when you fhall feeke
the centre in a fifeangle (whether it be to make a circle in it
other about it) you fhall drawe the two croffe lines, from one
angle to the other angle that lieth againfte it, and not to the
middle of any fide, as you did in the cinkangle.

THE XLVI. CONCLVSION.

To make a figure of fifteene equall fides
and angles in any circle appointed.

This rule is generall, that how many fides the figure fhall
haue

GEOMETRICALL.

haue, that ſhall be drawen in any circle, into ſo many partes iuſtely muſte the circles bee deuided. And therefore it is the more eaſier woorke commonly, to drawe a figure in a circle, then to make a circle in an other figure. Now therfore to end this concluſion, deuide the circle firſte into fiue partes, and then eche of them into thre partes againe: Or els firſt deuide it into thre partes, and then ech of thē into fiue other partes, as you liſt, and canne moſt readilye. Then draw lines betwene euery two prickes that be nigheſt togither, and ther wil appear rightly drawē the figure, of fiftene ſides, and angles equall. And ſo do with any other figure of what numbre of ſides ſo euer it bee.

FINIS.

THE SECOND BOOKE
OF THE PRINCIPLES

of Geometry, containing certaine
Theoremes, whiche may be cal=
led *Approued truthes. And be as
it were the mofte certaine
groundes, wheron the
practike cõclufions
of Geometry ar
founded.*

*Wherunto are annexed certaine declarations by
examples, for the right vnderftanding of the
fame, to the ende that the fimple reader
might not iuftly cõplain of hardnes
or obfcuritee, and for the fame
caufe ar the demonftra=
tions and iuft profes
omitted, vntill a
more conui
ent time.*

1551.

If truthe maie trie it selfe,
By Reasons prudent skyll,
If reason maie preuayle by right,
And rule the rage of will,
I dare the triall byde,
For truthe that I pretende.
And though some lyst at me repine,
Iuste truthe shall me defende.

THE PREFACE VNTO
the Theoremes.

Doubt not gentle reader, but as my argument is ſtraunge and vnacquainʒ ted with the vulgare tongue, ſo ſhall I of many men be ſtraungly talʒ ked of, and as ſtraunglye iudged. Some men will ſaye peraduenture, I mighte haue better imployed my tyme in ſome pleaſaunte hiſtorye, compriſinge matter of chiualrye. Some other wolde more haue preiſed my trauaile, if I hadde ſpente the like time in ſome morall matter, other in deciſing ſome controuerſy of religion. And yet ſome men (as I iudg) will not miſlike this kind of mater, but then will they wiſhe that I had vſed a more certaine order in placinge bothe the Propoſitions and Theoremes, and alſo a more exaꝗter proofe of eche of theim bothe, by demonſtrations mathematicall. Some alſo will miſlike my ſhortenes and ſimple plaineſſe, as other of other affeꝗtions diuerſely ſhall eſpye ſomwhat that they ſhall thinke blame worthy, and that miſſe ſomewhat, that thei wold wiſh to haue bene here vſed. So that euerie manne ſhall giue his verdiꝗte of me according to his phantaſie, vnto whome iointly, I make this my firſte anſwere : that as they ar many and in opinions verie diuers, ſo were it ſcarſe poſſi ble to pleaſe them all with anie one argumente, of what kinde ſo euer it were. And for my ſeconde annſwere, I ſaye thus. That is annye one argumente mighte pleaſe them all, then ſhoulde thei be thankfull vnto me for this kind of matter. For nother is there anie matter more ſtraunge in the engliſhe tungue, then this whereof neuer booke was written before now, in that tungue, and therefore oughte to delite all them, that deſire to vnderſtand ſtraunge matters, as moſt men commonlie doo. And againe the praꝗtiſe is ſo pleaſaunt in vʒ ſinge, and ſo profitable in appliynge, that who ſo euer dothe

a.ij.   deʒ

THE PREFACE.

*delite in anie of bothe, ought not of right to miſlike this arte.
And if any manne ſhall like the arte welle for it ſelf, but
ſhall miſlyke the fourme that I haue vſed in teachyng of it,
to hym I ſhall ſaie, Firſte, that I dooe wiſhe with hym that
ſome other man, whiche coulde better haue doone it, hadde
ſhewed his good will, and vſed his diligence in ſuche ſorte,
that I myght haue bene therby occaſioned iuſtely to haue left
of my laboure, or after my trauaile to haue ſuppreſſed my
bookes. But ſithe no manne hath yet attempted the like, as far
as I canne learne, I truſte all ſuche as bee not exerciſed in
the ſtudie of Geometrye, ſhall finde greate eaſe and fur=
theraunce by this ſimple, plaine, and eaſie forme of wri=
tinge. And ſhall perceaue the exacte woorkes of Theon,
and others that write on Euclide, a greate deale the ſom=
ner, by this blunte delineacion afore hande to them taughte.
For I dare preſuppoſe of them, that thing which I haue ſette
in my ſelfe, and haue marked in others, that is to ſaye, that
it is not eaſie for a man that ſhall trauaile in a ſtraunge arte,
to vnderſtand at the beginninge bothe the thing that is taught
and alſo the iuſte reaſon whie it is ſo. And by experience
of teachinge I haue tried it to bee true, for whenne I haue
taughte the propoſition, as it imported in meaninge, and
annexed the demonſtration withall, I didde perceaue that
it was a greate trouble and a painefull vexaction of mynde
to the learner, to comprehend bothe thoſe thinges at ones.
And therfore did I proue firſte to make them to vnderſtande
the ſence of the propoſitions, and then afterward did they
conceaue the demonſtrations muche ſoner, when they hadde
the ſentence of the propoſitions firſt ingrafted in their min=
des. This thinge cauſed me in bothe theſe bookes to o=
mitte the demonſtrations, and to vſe onlye a plaine forme
of declaration, which might beſt ſerue for the firſte intro=
duction. Whiche example hath beene vſed by other learned
menne before nowe, for not only Georgius Ioachimus Rhe=
ticus but alſo Boetius that wittye clarke did ſet forth ſome
whole books of Euclide, without any demonſtration or any*

other

THE PREFACE.

other declaratiõ at al. But & if I ſhal hereafter perceaue that it maie be a thankefull trauaile to ſette foorth the propoſitions of geometrie with demonſtrations, I will not refuſe to dooe it, and that with ſundry varietees of demonſtrations, bothe pleaſaunt and profitable alſo. And then will I in like ma= ner prepare to ſette foorth the other bookes, whiche now are lefte vnprinted, by occaſion not ſo muche of the charges in cuttyng of the figures, as for other iuſte hynderances, whiche I truſte hereafter ſhall bee remedied. In the meane ſeaſon if any man muſe why I haue ſette the Concluſions beefore the Theoremes, ſeynge many of the Theoremes ſeeme to include the cauſe of ſome of the concluſions, and therfore oughte to haue gone before them, as the cauſe goeth before the effeɛte. Here vnto I ſaie, that although the cauſe doo go beefore the effeɛt in order of nature, yet in order of teachyng the effeɛt muſt be fyrſt declared, and than the cauſe therof ſhewed, for ſo ſhal men beſt vnderſtãd things Firſt to lerne that ſuch thin= ges ar to be wrought, and ſecondarily what thei ar, and what thei do import, and thã thirdly what is the cauſe thereof. An other cauſe why y̆ the theoremes be put after the cõcluſions is this, whã I wrote theſe firſt cũcluſions (which was .iiij. yeres paſſed) I thought not then to haue added any theoremes, but next vnto y̆ cõcluſiõs to haue taught the order how to haue applied thē to work, for drawing of plottes, & ſuch like vſes. But afterward cõſidering the great cõmoditie y̆ thei ſerue for, and the light that thei do geue to all ſortes of praɛtiſe geo= metricall, beſyde other more notable benefites, whiche ſhall be declared more ſpecially in a place conuenient, I thoughte beſte to geue you ſome taſte of theym, and the pleaſaunt con= templation of ſuche geometrical propoſitions, which might ſerue diuerſelye in other bookes for the demonſtrations and proofes of all Geometricall woorkes. And in them, as well as in the propoſitions, I haue drawen in the Linearie examples many tymes more lynes, than be ſpoken of in the ex= plication of them, whiche is doone to this intent, that yf any manne lyſt to learne the demonſtrations by harte, as ſomme

a.iij.   lear=

THE PREFACE.

learned men haue iudged befte to doo) thofe fame men fhould finde the Linearye exaumples to ferue for this purpofe, and to wante no thyng needefull to the iufte proofe, whereby this booke maye bee wel approued to be more complete then many men wolde fuppofe it.

And thus for this tyme I wyll make an ende without any larger declaration of the commoditees of this arte, or any far ther anfweryng to that may bee obiected agaynft my hande= lyng of it, wyllyng them that myflike it, not to medle with it : and vnto thofe that will not difdaine the ftudie of it, I promife all fuche aide as I fhall be able to fhewe for their far= ther procedyng bothe in the fame, and in all other commodi= tees that thereof maie enfure. And for their incouragement I haue here annexed the names and brefe argumentes of fuche bookes, as I intende (God willyng) fhortly to fette forth, if I fhall perceaue that my paynes maie profyte other, as my de= fyre is.

The brefe argumentes of fuche bokes as ar appoyn ted fhortly to be fet forth by the author herof.

THE feconde part of Arithmetike, teachyng the wor= kyng by fractions, with extraction of rootes both fquare and cubike : And declaryng the rule of allegation, with fundrye pleafaunt exaumples in metalles and other thynges. Alfo the rule of falfe pofition, with dyuers examples not onely vul= gar, but fome appertaynyng to the rule of Algeber, applied vnto quantitees partly rationall, and partly furde.

THE arte of Meafuryng by the quadrate geometricall, and the diforders committed in vfyng the fame, not only re= ueled but reformed alfo (as muche as to the inftrument per= tayneth) by the deuife of a newe quadrate newely inuented by the author hereof.

THE arte of meafuryng by the aftronomers ftaffe, and by the aftronomers ryng, and the form of makyng them both.

THE arte of makyng of Dials, bothe for the daie and the nyght, with certayn new formes of fixed dialles for the moon
                                                                                    and

THE PREFACE.

and other for the sterres, whiche may bee sette in glasse win=
dowes, to serue by daie and by night. And howe you may by
those dialles knowe in what degree of the Zodiake not on=
ly the sonne, but also the moone is. And how many howrs old
she is. And also by the same dial to know whether any eclipse
shall be that moneth, of the sonne or of the moone.

The makyng and vse of an instrument, wherby you maye
not onely measure the distance at ones of all places that you
can see togyther, howe muche eche one is from you, and eue=
ry one from other, but also therby to drawe the plotte of a=
ny countreie that you shall come in, as iustely as maie be, by
mannes diligence and labour.

THE vse bothe of the Globe and the Sphere, and ther=
in also of the arte of Nauigation, and what instrumentes
serue beste thervnto, and of the trew latitude and longitude
of regions and townes.

Euclides woorkes in foure partes, with diuers demonstra=
tions Arithmeticall and Geometricall or Linearie. The fyrst
parte of platte formes. The second of numbres and quanti=
tees surde or irrationall. The third of bodies and solide for=
mes. The fourthe of perspectiue, and other thynges thereto
annexed.

BESIDE these I haue other sundrye woorkes partely
ended, And partely to bee ended, Of the peregrination of
man, and the originall of al nations, The state of tymes, and
mutations of realmes, The image of a perfect common welth,
with diuers other woorkes in naturall sciences, Of the won=
derfull workes and effectes in beastes, plantes, and minerals,
of whiche at this tyme, I will omitte the argumentes, bee=
cause thei doo appertaine littell to this arte, and handle other
matters in an other sorte.

> To haue, or leaue,
> Nowe maie you chuse,
> No paine to please,
> Willl I refuse.

The Theoremes of Geometry, before
WHICHE ARE SET FORTHE
certaine grauntable requeſtes
ᵂᵇⁱᶜᵇᵉ ſerue for demonſtrations
Mathematicall.

That frō any prickę to one other, there may   I.
be drawen a right line.

S for example A •————————• B.
A *being the one prickę, and* B. *the other,
you maye drawe betwene them from the
one to the other, that is to ſay, frome* A.
vnto B, *and from* B. *to* A.

That any right line of meaſurable length may   II.
be drawen forth longer, and ſtraight.

Example of A.B, *which as it is
a line of meaſurable lengthe, ſo*   A   B   C
may it be drawen forth farther, as for exaumple vnto C, *and
that in true ſtreightenes without crokinge.*

That vpon any centre, there   III.
*may be made a circle of anye
quātitee that a man wyll.*

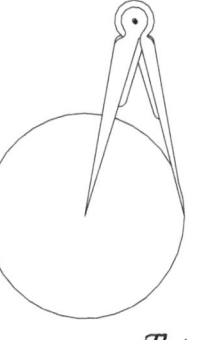

Let the centre be ſet to be A, *what ſhal
hinder a man to drawe a circle aboute
it, of what quantitee that he luſteth, as
you ſe the forme here : other bygger or
leſſe, as it ſhall lykę him to doo.*

b.i.   *That*

GRAVNTABLE

That all right angles be equall eche to other.

Set for an example A. and B, of which two
though A. ſeme the greatter angle to ſome
men of ſmall experience, it happeneth only
bicauſe that the lines aboute A, are lon=
ger thē the lines about B, as you may proue
by drawing them longer, for ſo ſhal B. ſeme
the greater angle yf you make his lines lon
ger then the lines that make the angle A. And to proue it by
demonſtration, I ſay thus. If any ij. right corners be not equal,
then one right corner is greater then an other, but that corner
which is greater then a right angle, is a blunt corner (by his de
finition) ſo muſt one corner be both a right corner and a blunt
corner alſo, which is not poſſible: And againe : the leſſer right
corner muſt be a ſharpe corner, by his definition, bicauſe it is
leſſe then a right angle. which thing is impoſſible. Therefore I
conclude that all right angles be equall.

Yf one right line do croſſe two other right
lines, and make ij. inner corners of one ſide leſ
ſer thē ij. righte corners, it is certaine, that if
thoſe two lines be drawen forth right on that
ſide that the ſharpe inner corners be, they wil
at lēgth mete togither, and croſſe on an other.

The ij. lines beinge as
A.B. and C.D, and the
third line croſſing them
as dooth heere E.F, ma=
king ij. inner cornes (as
ar G.H.) leſſer then two
right corners, ſith ech of

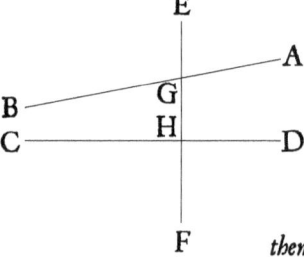

them

REQVESTES.

them is leſſe then a right corner, as your eyes maye iudge, then ſay I, if thoſe ij. lines A.B. *and* C.D. *be drawen in lengthe on that ſide that* G. *and* H. *are, they will at length meet and croſſe one another.*

Two right lines make no platte forme.

A platte forme, as you harde before, hath bothe length and bredthe, and is incloſed with lines as with his boundes, but ij. right lines cannot incloſe al the bon= des of any platte forme. Take for an ex= ample firſte theſe two right lines A B. *and* A.C, *whiche meete togither in* A, *but yet cannot be called a platte forme, bicauſe there is no bond from* B. *to* C, *but if you will drawe a line betwene them twoo, that is frome* B. *to* C, *then will it be a platte forme, that is to ſay, a triangle, but then are there iij. lines, and not only ij. Likewiſe may you ſay of* D.E. *and* F.G, *whiche doo make a platte forme, nother yet can they make any without helpe of two lines more, whereof the one muſt be drawen from* D. *to* F, *and the other frome* E. *to* G, *and then will it be a longe ſquare. So then of two right lines can bee made no platte forme. But of ij. croked lines be made a platte forme, as you ſe in the eye form. And alſo of one right line, & one cro ked line, maye a platte fourme bee made, as the ſemicircle* F. *doothe ſette forth.*

b.ij.   *Certayne*

COMMON.

Certayn common sentences manifest to sence, and acknowledged of all men.

The firste common sentence.

W*hat so euer things be equal to one other thinge, those same bee equall betwene them selues.*

Examples therof you may take both in greatnes and also in numbre. First (though it pertaine not proprely to geometry, but to helpe the vnderstandinge of the rules, whiche may bee wrought by bothe artes) thus may you perceaue. If the summe of monnye in my purse, and the mony in your purse be equall eche of them to the mony that any other man hathe, then must needes your mo=

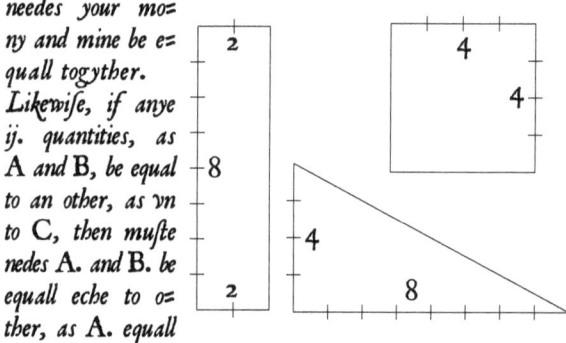

ny and mine be e= quall togyther. Likewise, if anye ij. quantities, as A *and* B, *be equal to an other, as vn to* C, *then muste nedes* A. *and* B. *be equall eche to o= ther, as* A. *equall to* B, *and* B. *equall to* A, *whiche thinge the better to peceaue, tourne these quantities into numbre, so shall* A. *and* B. *make sixteene, and* C. *as many. As you may perceaue by multiplyng the numbre of their sides togither.*

The seconde common sentence.

And if you adde equall portions to thin= ges that be equall, what so amounteth of them shal

SENTENSES.

ſhall be equall.

Example, If *you and I haue like ſummes of mony, and then receaue eche of vs like ſummes more, then our ſummes wil be like ſtyll.* Alſo *if* A. *and* B. *(as in the former example) bee e= quall, then by adding an equal portion to them both, as to ech of them, the quarter of* A. *(that is foure) they will be equall ſtill.*

The thirde common ſentence.

And if you abate euen portions from things that are equal, thoſe parts that remain ſhall be equall alſo.

This you may perceaue by the laſte example. For that that was added there, is ſubtraƈted heere, and ſo the one doothe approue the other.

The fourth common ſentence.

If you abate equalle partes from vnequal thin ges, the remainers ſhall be vnequall.

As bicauſe that a hundreth and eight and forty be vnequal if I take tenne from them both, there will remaine nynetye and eight and thirty, which are alſo vnequall, and likewiſe in quantities it is to be iudged.

The fifte common ſentence.

When euen portions are added to vnequalle thinges, thoſe that amounte ſhalbe vnequall.

b.iij.   So if

COMMON.

*So if you adde twenty to fifty, and lyke ways to nynty, you
shall make seuenty and a hundred and ten whiche are no lesse
vnequall, than were fifty and nynty.*

The fyxt common sentence.

If two thinges be double to any other, those same two thinges are equal togither.

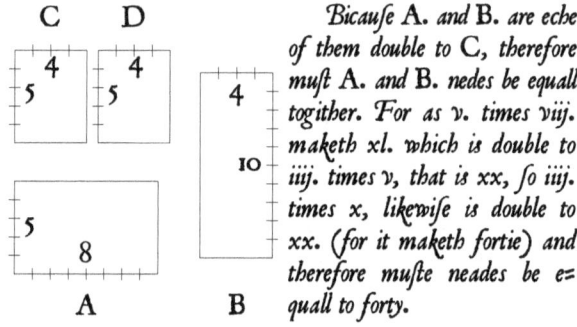

C   D   *Bicause* A. *and* B. *are eche
of them double to* C, *therefore
must* A. *and* B. *nedes be equall
togither. For as* v. *times* viij.
maketh xl. *which is double to*
iiij. *times* v, *that is* xx, *so* iiij.
times x, *likewise is double to*
xx. (*for it maketh fortie*) *and
therefore muste neades be e=
quall to forty.*

A   B

The feuenth common sentence.

If any two thinges be the halfes of one other thing, than are thei .ij. equall togither.

So are D. *and* C. *in the laste example equal togyther, bicause
they are eche of them the halfe of* A. *other of* B, *as their num
bre declareth.*

The eyght common sentence.

If any one quantitee be laide on an o= ther, and thei agree, so that the one

                                                        excedeth

SENTENSES.

excedeth not the other, then are they e=
quall togither.

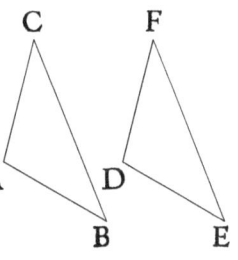

As if this figure A.B.C, *be layed on*
that other D.E.F, *so that* A. *be layed to*
D, B. *to* E, *and* C. *to* F, *you shall see them*
agre in sides exactlye and the one not to
excede the other, for the line A.B. *is e=*
quall to D.E, *and the third lyne* C.A, *is*
equal to F.D *so that euery side in the one*
is equall to some one side of the other.
Wherfore it is playne, that the two triangles are equall to=
gither.

The nynth common sentence.

Euery whole thing is greater than any
of his partes.

This sentence nedeth none example. For the thyng is more
playner then any declaration, yet considering that other com=
men sentence that foloweth nexte that.

The tenthe common sentence.

Euery whole thinge is equall to all his
partes taken togither.

It shall be mete to expresse both ẘ one example, for of thys
last sētence many mē at the first hearing do make a doubt. Ther
fore as in this example of the circle deuided into sūdry partes
it

COMMON.

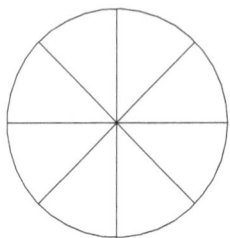

it doeth appere that no parte can be fo greate as the whole circle, (accordyng to the meanyng of the eight fentence) fo yet it is certain, that all thofe eight par= tes together be equall vnto the whole circle. And this is the meanyng of that common fentence (whiche many vfe, and fewe do rightly vnderftand) that is,

that All the partes of any thing are nothing els, but the whole. *And contrary waies:* The whole is nothing els, but all his partes taken togither. *whiche faiynges fome haue vnderftand to meane thus: that all the partes are of the fame kind that the whole thyng is: but that that meanyng is falfe, it doth plainly appere by this figure* A. B, *whofe partes* A. *and* B, *are trian= gles, and the whole figure is a fquare, and fo are they not of one kind. But and if they applie it to the matter or fubftance of thin ges (as fome do) then is it mofte falfe, for e=*

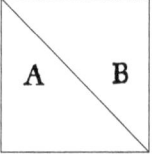

uery compound thyng is made of partes of diuerfe matter and fubftance. Take for example a man, a houfe, a boke, and all o= ther compound thinges. Some vnderftand it thus, that the par= tes all together can make none other forme, but that that the whole doth fhewe, whiche is alfo falfe, for I maie make fiue hundred diuerfe figures of the partes of fome one figure, as you fhall better perceiue in the third boke. And in the meane feafõ take for an exãple this fquare figure folowing A.B.C.D, *ᵂ is deuided but into two parts, and yet (as you fe) I haue made fiue figures more befide the firfte, with onely diuerfe ioynyng of thofe two partes. But of this fhall I fpeake more largely in an other place, in the mean feafon content your felf with thefe principles, whiche are certain of the chiefe groundes wheron all demonftrations mathematical are fourmed, of which though the mofte parte feeme fo plaine, that no childe doth doubte of them, thinke not therfore that the art vnto whiche they ferue, is fimple, other childifhe, but rather confider, howe certayne the*

GEOMETRICALL.

the profes of
that arte is, y
hath for his gro
ūdes soche pla=
yne truthes, &
as I may say,
suche vndow=
btfull and senf
ible principles,
And this is the
cause why all
learned menne
dooth approue
the certenty of
geometry, and
cōsequently of
the other artes
mathematical,
which haue the
grounds (as A=
rithmetike, mu
sike and astro=

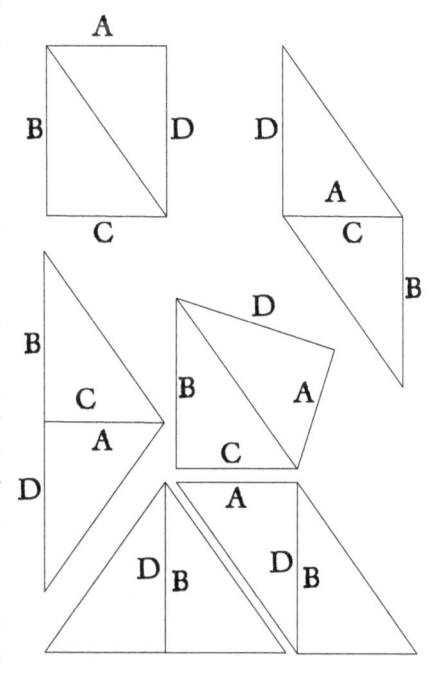

nomy) aboue all other artes and sciences, that be vsed amōgest
men. Thus muche haue I sayd of the first principles, and now
will I go on with the theoremes, whiche I do only by exam=
ples declar, minding to reserue the proofes to a peculiar boke
which I will then set forth, when I perceaue this to be thank=
fully taken of the readers of it.

The theoremes of Geometry brieflye
declared by shorte examples.

The firste Theoreme.

W hen ij. triangles be so drawen, that the
one of thē hath ij. sides equal to ij. sides of the
other

THEOREMES

other triangle, and that the angles encloſed
with thoſe ſides, bee equal alſo in bothe trian=
gles, then is the thirde ſide likewiſe equall in
them.   And the whole triangles be of one
greatnes, and euery angle in the one equall to
his matche angle in the other, I meane thoſe
angles that be incloſed with like ſides.

Example.

This triangle A.B.C. hath ij.
ſides (that is to ſay) C.A. and
C.B, equal to ij. ſides of the
other triangle F.G.H, for A.
C. is equall to F.G, and B.C.
is equall to G.H. And alſo
the angle C. contayned bee=
tweene F.G, and G.H, for
both of them anſwere to the
eight parte of a circle. Ther
fore doth it remayne that A.
B. whiche is the thirde lyne
in the firſte triangle, doth a=
gre in lengthe
with F.H, w is the

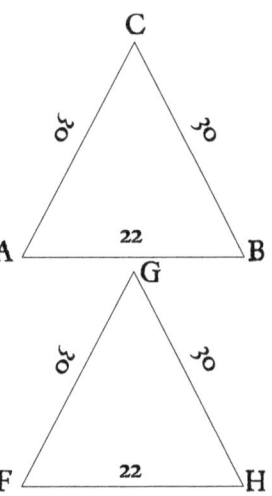

third line in y ſecod tri agle & y hole triagle.
A.B.C. muſt nedes be equal to y hole tri angle F.G.H. And
euery corner equall to his match, that is to ſay, A. equall to F,
B. to H, and C. to G, for thoſe bee called match corners,
which are incloſed with like ſides, other els do lye againſt like
ſides.

The ſecond Theoreme.

In twileke triangles the ij. corners that be

GEOMETRICALL.

about the grōud line, are equal togither. And if the fides that be equal, be drawē out in lēgth thē wil the corners that are vnder the ground line, be equal alfo togither.

Example

A.B.C. *is a twileke triangle, for the one fide* A.C, *is equal to the other fide* B.C. *And therfor I faye that the inner corners* A. *and* B, *which are about the ground lines, (that is* A.B.) *be equall togither. And farther if* C.A. *and* C.B. *bee drawen forthe vnto* D *and* E. *as you fe that I haue drawen them, then faye I that the two vtter angles vnder* A. *and* B, *are equal alfo togither : as the theorem faid. The profe wherof, as of al the reft, fhal apeare in Euclide, whome I intende to fet foorth in englifh with fondry new additions, if I may perceaue that it wil be thankfully taken.*

The thirde Theoreme.

If in annye triangle there bee twoo angles equall togither, then fhall the fides, that lie against thofe angles, be equal alfo.

Example

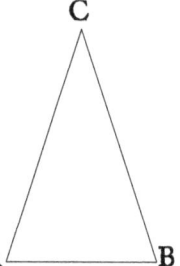

This triangle A.B.C. *hath two corners equal eche to other, that is* A. *and* B, *as I do by fuppofition limite, wherfore it foloweth that the fide* A.C, *is equal to that other side* B.C, *for the fide* A.C, *lieth against the angle* B, *and the fide* B.C, *lieth against the angle* A.

c.ij.   *The*

THEOREMES

The fourth Theoreme.

When two lines are drawen frō the endes of anie one line, and meet in anie pointe, it is not poſſible to draw two other lines of like lengthe, ech to his match that ſhal begī at the ſame poin tes, and end in anie other pointe then the twoo firſt did.

Example.

The firſt line is A.B, *on which I haue erected two other lines* A.C, *and* B. C, *that meete in the pricke* C, *where= fore I ſay, it is not poſſible to draw ij. other lines from* A. *and* B. *which ſhal mete in one point (as you ſe* A.D. *and* B.D. *mete in* D) *but that the match li nes ſhal be vnequal, I mean by* match lines, *the two lines on one ſide that is the ij. on the right hand, or the ij. on the lefte hand, for as you ſe in this ex ample* A.D. *is longer thē* A.C, *and* B.C. *is longer then* B.D. *And it is not poſſible, that* A.C. *and* A.D. *ſhall bee of one lengthe, if* B.D. *and* B.C. *bee like longe. For if one couple of matche lines be equall (as the ſame example* A.E. *is equall to* A.C. *in length) then muſt* B.E. *needes be vnequall to* B.C. *as you ſee, it is here ſhorter.*

The fifte Theoreme.

If two triāgles haue there ij. ſides equal one to an other, and their groūd lines equal alſo, then ſhall

GEOMETRICALL.

ſhall their corners, whiche are contained be=
twene like ſides, be equall one to the other.

Example.

Becauſe theſe two triangles A.B.C, *and* D.E.F. *haue*
two ſides equall one to an other.
For A.C. *is equall to* D.F,
and B.C. *is equall to* E.F, *and*
again their groūd lines A.B.
and D.E. *are lyke in length,*
therfore is eche angle of the
one triangle equall to ech an
gle of the other, comparyng
together thoſe angles that are contained within lyke ſides, ſo is
A. *equall to* D, B. *to* E, *and* C. *to* F, *for they are contayned*
within like ſides, as before is ſaid.

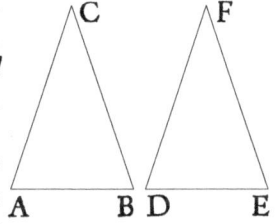

The ſixt Theoreme.

When any right line ſtandeth on an other, the
ij. angles that thei make, other are both right
angles, or els equall to .ij. righte angles.

Example.

A.B. *is a right line, and on it*
there doth light another right
line, drawen from C. *perpen=*
dicularly on it, therefore ſaie
I, that the .ij. angles that thei
do make, are .ij. right angles
as maie be iudged by the defi=
nition of a right angle. But in
the ſecond part of the exam=
ple, where A.B. *beyng ſtill the right line, on whiche* D. *ſtan=*

deth

THEOREMES

deth in flope wayes, the two angles that be made of them are
not righte angles, but yet they are equall to two righte angles,
for fo muche as the one is to greate, more then a righte angle,
fo muche iufte is the other to little, fo that bothe togither are
equall to two right angles, as you maye perceiue.

The feuenth Theoreme.

If .ij. lines be drawen to any one pricke in an
other lyne, and thofe .ij. lines do make with the
fyrft lyne, two right angles, other fuche as be
equall to two right angles, and that towarde
one hande, than thofe two lines doo make one
ftreyght lyne.

Example.

A.B. *is a ftreyght lyne,*
on which there doth lyght
two other lines one frome
D, *and the other frome* **C,**
but confiderynge that they
meete in one pricke **E,** *and*
that the angles on one hand
be equal to two right cor=

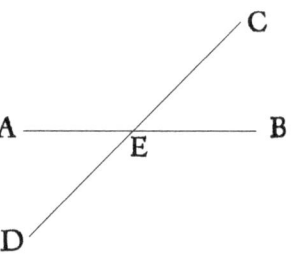

ners (as the lafte theoreme dothe declare) therfor maye **D.E.**
and **E.C.** *be counted for one ryght lyne.*

The eight Theoreme.

When two right lines do cut one an other croffeways
they do make their matche angles equall.

Ex=

GEOMETRICALL.

Example.

What matche angles are, I haue tolde you in the defini= tions of the termes. And here A, *and* B. *are matche corners in this example, as are alſo* C. *and* D, *ſo that the corner* A, *is equall to* B, *and the angle* C, *is equall to* D.

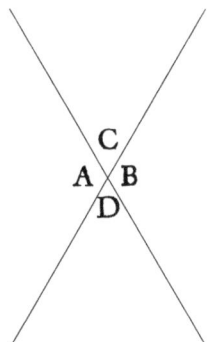

The nynth Theoreme.

Whan ſo euer in any triangle the line of one ſide is drawen forthe in lengthe, that vtter an= gle is greater than any of the two inner cor= ners, that ioyne not with it.

Example.

The triangle A.D.C *hathe hys grounde lyne* A.C. *drawen forthe in lengthe vnto* B, *ſo that the vtter corner that it maketh at* C, *is greater then any of the two in= ner corners that lye a= gainſte it, and ioyne not wyth it, whyche are* A.

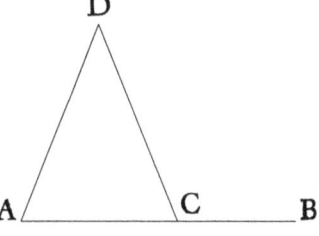

and D, *for they both are leſſer then a ryght angle, and be ſharpe angles, but* C. *is a blonte angle, and therfore greater then a ryght angle.*

The tenth Theoreme.

In euery triangle any .ij. corners, how ſoe= uer you take thē, ar leſſe thē ij. right corners.

Example.

THEOREMES

Example.

In the firste triangle E, whiche is a threlyke, and therfore hath all his an= gles sharpe, take anie twoo corners that you will, and you shall perceiue that they be lesser then .ij. right cor= ners, for in euery triangle that hath all sharpe corners (as you see it to be in this example) euery corner is lesse then a right corner. And therfore al= so euery two corners must nedes be lesse then two right corners. Fur= thermore in that other triangle mar= ked with M, whiche hath .ij. sharpe corners and one right, any .ij. of them also are lesse then two right angles. For though you take the right corner for one, yet the other whiche is a sharpe corner, is lesse then a right corner. And so it is true in all kindes of tri= angles, as you maie perceiue more plainly by the .xxij. Theo= reme.

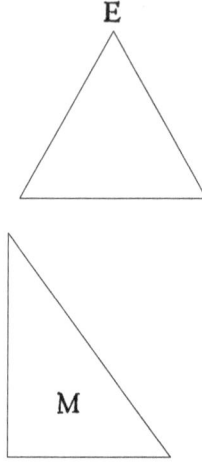

The .xi. Theoreme.

In euery triangle, the greattest side lieth againft the greattest angle.

Example.

As in this triangle A.B.C, the greattest angle is C. And A.B. (whiche is the side that lieth againft it) is the greatest and longest side. And contra= ry waies, as A.C. is the shor= test side, so B. (whiche is the angle liyng againft it) is the smallest

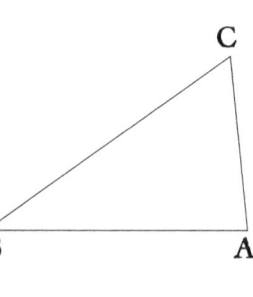

GEOMETRICALL.

ſmalleſt and ſharpeſt angle, for this doth folow alſo, that as the longeſt ſide lyeth againſt the greateſt angle, ſo it that foloweth

The twelft Theoreme.

In euery triangle the greatteſt angle lieth againſt the longeſt ſide.

For theſe ij. theoremes are one in truthe.

The thirtenth theoreme.

In euerie triangle anie ij. ſides togither how ſo euer you take them, are longer thē the thirde.

A

For example you ſhal take this triangle A.B.C. *which hath a vee ry blunt corner, and therfore one of his ſides greater a good deale then any of the other, and yet the ij. leſſer ſides togither ar greate then it. And if it bee ſo in a blunte angeled triangle, it muſt nedes be true in all other, for there is no other kinde of triangles that hathe the one ſide ſo greate aboue the other ſids, as thei y haue blunt corners.*

B   C

The fourtenth theoreme.

If there be drawen from the endes of anie ſide of a triangle .ij. lines metinge within the triangle, thoſe two lines ſhall be leſſe then the other twoo ſides of the triangle, but yet the

THEOREMES

corner that thei make, ſhall bee greater then that corner of the triangle, whiche ſtandeth ouer it.

Example.

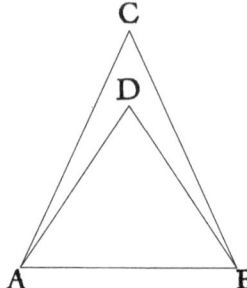

A.B.C. *is a triangle on whoſe ground line* A.B. *there is dra wen ij. lines, from the ij. endes of it, I ſay from* A. *and* B, *and they meete within the trian= gle in the pointe* D, *wherfore I ſay, that as thoſe two lynes* A.D. *and* B.D, *are leſſer then* A.C. *and* B.C, *ſo the angle* D, *is greatter then the angle* C, *which is the angle againſt it.*

The fiftenth Theoreme.

If a triangle haue two ſides equall to the two ſides of an other triangle, but yet the ãgle that is contained betwene thoſe ſides, greater then the like angle in the other triangle, then is his grounde line greater then the grounde line of the other triangle.

Example.

A.B.C. *is a triangle, whoſe ſides* A.C. *and* B.C, *are equall to* E.D. *and* D.F, *the two ſides of the triangle* D.E.F, *but bicauſe the angle in* D, *is greatter then the angle* C. *whiche are the ij. an= gles contayned betwene the equal ly=* *nes*

GEOMETRICALL.

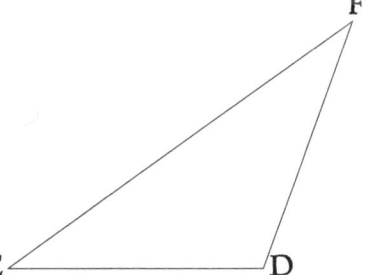

nes) *therfore mufte the ground line* E.F. *nedes bee greatter thenne the grounde line* A.B, *as you fe plainely.*

The xvi. Theoreme.

If a triangle haue twoo fides equalle to the two fides of another triangle, but yet hathe a longer ground line thē that other triangle, then is his angle that lieth betwene the equall fides, greater thē the like corner in the other triangle.

Example.

This Theoreme is nothing els, but the fentence of the laft Theoreme turned backward, and therfore nedeth none other profe nother declaration, then the other example.

The feuententh Theoreme.

If two triangles be of fuch fort, that two angles of the one be equal to ij. angles of the o= ther, and that one fide of the one be equal to on fide of the other, whether that fide do adioyne to one of the equall corners, or els lye againfte

THEOREMES

one of them, then ſhall the other twoo ſides of thoſe triangles bee equalle togither, and the thirde corner alſo ſhall be equall in thoſe two triangles.

Example.

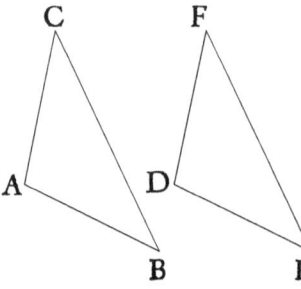

Bicauſe that A.B.C, *the one triangle hath two cor ners* A. *and* B, *equal to* D. E, *that are twoo corners of the other triangle.* D. E.F. *and that they haue one ſide in theym bothe e= quall, that is* A.B, *which is equal to* D.E, *therefore* ſhall both the other ij. ſides be equall one to an other, as A.C. and B.C. equall to D.F and E.F, and alſo the thirde angle in them both ſhal be equall, that is, the angle C. ſhal be equall to the angle F.

The eightenth Theoreme.

When on .ij. right lines ther is drawen a third right line croſs waies, and maketh .ij. matche corners of the one line equall to the like twoo matche corners of the other line, then ar thoſe two lines gemmow lines, or paralleles.

Example.

The

GEOMETRICALL.

The .ij. fyrſt lynes are A.
B. and C.D, the thyrd lyne
that croſſeth them is E.F. A _____ E /M ____ B
And bycauſe that E.F. ma=   N /G
keth ij. matche angles with   K /L
A.B, equall to .ij. other lyke C   H /F   D
matche angles on C.D, (that
is to ſay E.G, equall to K.
F, and M.N. equall alſo to H,L.) therfore are thoſe ij. lynes
A.B. and C.D. gemow lynes, vnderſtand here by lyke mat=
che corners, thoſe that go one way as doth E.G, and K.F,
lykeways N.M, and H.L, for as E.G. and H.L, other N.M.
and K.F. go not one waie, ſo be not they lyke match corners.

The nyntenth Theoreme.

When on two right lines there is drawen a
thirde right line croſſewaies, and maketh the
ij. ouer corners towarde one hande equall to=
gither, then ar thoſe .ij. lines paralleles. And
in like maner if two inner corners toward one
hande, be equall to .ij. right angles.

Example.

As the Theoreme dothe ſpeake of .ij. ouer angles, ſo muſte
you vnderſtande alſo of .ij. nether angles, for the iudgement is
lyke in bothe. Take for an example the figure of the laſt theo=
reme, where A.B, and C.D, be called paralleles alſo, bicauſe
E. and K, (whiche are .ij. ouer corners) are equall, and lyke
waies L. and M. And ſo are in lyke maner the nether corners
N. and H, and G. and F. Nowe to the ſeconde parte of the
theoreme, thoſe .ij. lynes A.B. and C.D, ſhall be called pa=
ralleles, becauſe the ij. inner corners. As for example thoſe
two that bee toward the right hande (that is G. and L.) are e=

THEOREMES

*quall (by the fyrſt parte of this nyntenth theoreme) therfore
muſte* G. *and* L. *be equall to two ryght angles.*

The xx. Theoreme.

*When a right line is drawen croſſe ouer .ij.
right gemow lines, it maketh .ij. match corners of the one line, equall to two matche corners of the other line, and alſo bothe ouer cor
ners of one hande equall togither, and bothe
nether corners likewaies, and more ouer two
inner corners, and two vtter corners alſo towarde one hande, equall to two right angles.*

Example.

Bycauſe A.B. *and* C.D, *(in the laſte figure) are paralleles,
therefore the two matche corners of the one lyne, as* E.G. *be
equall vnto the .ij. matche corners of the other line, that is*
K.F, *lykewaies* M.N, *equall to* H.L. *And alſo* E. *and*
K. *bothe ouer corners of the lefte hande equall togyther,
and ſo are* M. *and* L, *the two ouer corners on the ryghte
hande, in lyke maner* N. *and* H. *the two nether corners on
the lefte hande, equall eche to other, and* G. *and* F. *the two
nether angles on the right hande equall togither.*
Farthermore yet G. *and* L. *the .ij. inner angles on the right
hande bee equall to two right angles, and ſo are* M. *and* F.
*the .ij. vtter angles on the ſame hande, in lyke manner ſhall
you ſay of* N. *and* K. *the two inner corners on the left hand.
and of* E. *and* H. *the two vtter corners on the ſame hande.
And thus you ſee the agreable ſentence of theſe .iij. theoremes to tende to this purpoſe, to declare by the angles how to
iudge paralleles, and contrary waies howe you may by paralleles iudge the proportion of the angles.*

The

GEOMETRICALL.

The xxi. Theoreme.

What ſo euer lines be paralleles to any other line, thoſe ſame be paralleles togither.

Example.

A.B. *is a gemow line, or a paral=*
lele vnto C.D. *And* E.F, *lykewaies*
is a parallele vnto C.D. *Wherfore it*
foloweth, that A.B. *muſt nedes bee a parallele vnto* E.F.

```
A————————————B
C————————————C
E————————————F
```

The .xxij. theoreme.

In euery triangle, when any ſide is drawen forth in length, the vtter angle is equall to the ij. inner angles that lie againſte it. And all iij. inner angles of any triangle are equall to ij. right angles.

Example.

The triangle beeyng
A.D.E. *and the ſyde* A.
E. *drafwen foorthe vnto*
B, *there is made an vtter*
corner, whiche is C, *and*
this vtter corner C, *is*
equall to bothe the in=
ner corners that lye a=

gaynſt it, whyche are A. *and* D. *And all thre inner corners,*
that is to ſay, A.D. *and* E, *are equall to two ryght corners,*
whereof it foloweth, that all the three corners of a=
ny one triangle are equall to all the three corners
of euerye other triangle. *For what ſo euer thynges*
are equalle to anny one thyrde thynge, thoſe ſame are
                                                              equall

THEOREMES

equalle togitther, by the fyrste common sentence, so that
bycause all the .iij. angles of euery triangle are equall to
two ryghte angles, and all ryghte angles bee equall togy=
ther (by the fourth requeſt) therfore muſt it nedes folow, that
all the thre corners of euery triangle (accomptyng them to=
gyther) are equall to iij. corners of any other triangle, taken
all togyther.

The .xxiii. theoreme.

*When any ij. right lines doth touche and cou=
ple .ij. other righte lines, whiche are equall in
length and paralleles, and if thoſe .ij. lines bee
drawen towarde one hande, then are thei alſo
equall together, and paralleles.*

Example.

A.B. and C.D. are ij. ryght
lynes and paralleles, and e=
quall in length, and they ar
touched and ioyned togither
by ij. other lynes A.C. and
B.D, this beyng ſo, and A.C.
and B.D. beyng drawen to=
warde one ſyde (that is to
ſaye, bothe towarde the lefte hande) therefore are A.C. and
B.D. bothe equall and alſo paralleles.

The .xxiiii. theoreme.

*In any likeiamme the two contrary ſides ar
equall togither, and ſo are eche .ij. contrary
angles, and the bias line that is drawen in it,
dothe diuide it into two equall portions.*

Exam=

GEOMETRICALL.

Example.

A   B   C   D

E   F   G   H

Here ar two likeiammes ioyned togither, the one is a longe square A.B.E, and the other is a losengelike D.C. E.F. which ij. likeiammes ar proued equall togither, by= cause they haue one ground line, that is, F.E, And are made betwene one payre of gemow lines, I meane A.D. and E.H. By this Theoreme may you know the arte of the righte measuringe of likeiammes, as in my booke of measuring I wil more plainly declare.

The xxvi. Theoreme.

All likeiammes that haue equal grounde lines and are drawen betwene one paire of pa= ralleles, are equal togither.

Example.

Fyrste you muste marke the difference betwene this Theo= reme and the laste, for the laste Theoreme presupposed to the diuers likeiammes one ground line common to them, but this theoreme doth presuppose a diuers ground line for euery like iamme, only meaning them to be equal in length, though they be diuers in numbre. As for example. In the last figure ther are two parallels, A.D. and E.H, and betwene them are drawen thre likeiammes, the firste is, A.B.E.F, the seconde is E.C.D.F, and the thirde is C.G.H.D. The firste and the seconde haue one ground line, (that is E.F.) and therfore in so muche as they are betwene one paire of paralleles, they are equall accordinge to the fiue and twentye Theoreme, but the thirde likeiamme that is C.G.H.D. hathe his grounde line G.H, seuerall frome

e.i.   the

THEOREMES

*the other, but yet equall vnto it. wherefore the third likeiame
is equall to the other two firſte likeiammes. And for a proofe
that* G.H. *being the grōud line of the third likeiamme, is equal
to* E.F, *whiche is the ground line to both the other likeiames,
that may be thus declared,* G.H. *is equall to* C.D. *ſeynge they
are the contrary ſides of one likeiamme (by the foure and twē
ty theoreme) and ſo are* C.D. *and* E.F. *by the ſame theoreme.
Therfore ſeynge both thoſe ground lines* F.E. *and* G.H, *are e=
quall to one thirde line (that is* C.D.) *they muſt nedes bee e=
quall togyther by the firſte common ſentence.*

The xxvii. Theoreme.

*All triangles hauinge one grounde lyne,
an ſtanding betwene one paire of parallels, ar
equall togither.*

Example.

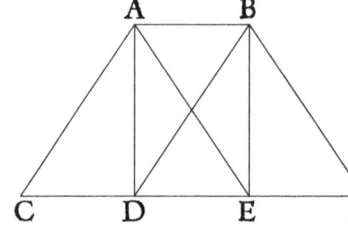

A.B. *and* C.F. *are twoo
gemowe lines, betweene
which there be made two tri
angles,* A.D.E. *and* D.E.B.
ſo that D.E, *is the common
ground line to them bothe.
Wherfore it doth folow, that
thoſe two triangles* A.D.E.
and D.E.B. *are equall eche to other.*

The xxviii. Theoreme.

*All traingles that haue like long ground
lines, and bee made betweene one paire of ge=
mow lines, are equall togither.*

                                                        Ex=

GEOMETRICALL.

Example

Example of this Theoreme you may fee in the laft figure,
where as fixe triangles made betwene thofe two gemowe li=
nes A.B. *and* C.F, *the firft triangle is* A.C.D, *the feconde*
is A.D.E, *the thirde is* A.D.B, *the fourth is* A.B.E, *the fifte*
is D.E.B, *and the fixte is* B.E.F, *of which fixe triangles,* A.
D.E. *and* D.E.B. *are equall, bicaufe they haue one common*
grounde line. And fo likewife A.B.E. *and* A.B.D, *whofe com*
men grounde line is A.B, *but* A.C.D. *is equal to* B.E.F, *being*
both betwene one couple of paraellels, not bicaufe thei haue
one ground line, but bicaufe they haue their ground lines e=
quall, for C.D. *is equall to* E.F, *as you may declare thus.* C.D,
is equall to A.B. *(by the foure and twenty Theoreme) for thei*
are two contrary fides of one lykeiamme. A.C.D.B, *and* E.F
by the fame theoreme, is equall to A.B, *for thei ar the two y̆*
contrary fides of the likeiamme, A.E.F.B, *wherfore* C.D. *muft*
needes be equall to E.F. *likewife the triangle* A.C.D, *is equal*
to A.B.E, *bicaufe they ar made betwene one paire of parallels*
and haue their groundlines like, I meane C.D. *and* A.B. *A=*
gaine A.D.E, *is equal to eche of them both, for his ground line*
D.E, *is equall to* A.B, *in fo muche as they are the contrary fi=*
des of one likeiamme, that is the long fquare A.B.D.E. *And*
thus may you proue the equalnes of all the refte.

The xxix. Theoreme.

Al equal triangles that are made on one
grounde line, and rife one waye, muft needes
be betwene one paire of parallels.

Example.

Take for example A.D.E, *and* D.E.B, *which as the xxvii.*

THEOREMES

conclufion dooth proue) *are equall togither, and as you fee,
they haue on ground line* D.E. *And againe they rife towarde
one fide, that is to fay, vpwarde toward the line* A.B, *wher
fore they muft needes be inclofed betweene one paire of pa=
rallels, which are heere in this example* A.B. *and* D.E.

The thirty Theoreme.

*Equal triangles that haue their ground lines
equal, and be drawē toward one fide, ar made
betwene one paire of paralleles.*

Example.

*The example that declared the laft theoreme, maye well
ferue to the declaracion of this alfo. For thofe* ij. *theoremes do
diffre but in this one pointe, that the lafte theoreme meaneth
of triangles, that haue one ground line common to them both,
and this theoreme dothe prefuppofe the grounde lines to bee
diuers, but yet of one length, as* A.C.D, *and* B.E.F, *as they
are* ij. *equall triangles approued, by the eighte and twentye
Theorem, fo in the fame Theorem it is declared, ꝙ their groūd
lines are equall togither, that is* C.D, *and* E.F, *now this bee=
ynge true, and confidering that they are made towarde one
fide, it foloweth, that they are made betwene one paire of pa
rallels when I faye, drawen towarde one fide, I meane that
the triangles muft be drawen other both vpward frome one
parallel, other els both downward, for if the one be drawen
vpward and the other downward, then are they drawen be=
twene two paire of parallels, prefuppofinge one to bee dra=
wen by their groundline, and then do they ryfe toward con
trary fides.*

The

GEOMETRICALL.

The xxxi. theoreme.

If a likeiamme haue one ground line with a triangle, and be drawen betwene one paire of paralleles, then ſhall the likeiamme be double to the triangle.

Example.

A.H. and B.G, are .ij. ge= mow lines, betwene which there is made a triangle B.C G, and a lykeiamme, A.B.G. C, whiche haue a grounde lyne, that is to ſaye, B.G. Therfore doth it folow that the lykeiamme A.B.G.C. is

double to the triangle B.C.G. For euery halfe of that lyke= iamme is equall to the triangle, I meane A.B.F.E. other F.E. C.G. as you may conieƈlure by the .xv. concluſion geometrical.

And as this Theoreme dothe ſpeake of a triangle and like iamme that haue one groundelyne, ſo is it true alſo, yf theyr groundelynes bee equall, though they bee dyuers, ſo that thei be made betwene one payre of paralleles. And hereof may you perceaue the reaſon, why in meaſuryng the platte of a triangle, you muſt multiply the perpendicular lyne by halfe the grounde lyne, or els the hole grounde lyne by halfe the perpendicular, for by any of theſe bothe waies is there made a lykeiamme equall to halfe ſuche a one as ſhulde be made on the ſame hole grounde lyne with the triangle, and betweene one payre of paralleles. Therfore as that lykeiamme is dou= ble to the triangle, ſo the halfe of it, muſt needes be equall to the triangle. Compare the .xv. concluſion with this theoreme.

The .xxxii. Theoreme.

In all likeiammes where there are more than

one

THEOREMES

one made aboute one bias line, the fill ſquares
of euery of them muſt nedes be equall.

Example.

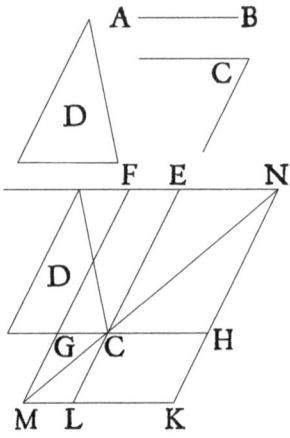

Bias lyne.

Fyrſt before I declare the ex=
amples, it ſhal be mete to ſhew
the true vnderſtādyng of this
theorem. Therfore by the Bi=
as line, *I meane that lyne,*
whiche in any ſquare figure
dooth runne from corner to
corner. And euery ſquare
which is diuided by that bias
line into equall halues from
corner to corner (that is to
ſay, into .ij. equall triangles)
thoſe be counted to ſtande
aboute one bias line, *and*
the other ſquares, whiche touche that bias line, with one
of their corners onely, thoſe doo I call Fyll ſquares, ac=
cordyng to the greke name, whiche is anapleromata, and
called in latin ſupplementa, bycauſe that they make one ge=
nerall ſquare, includyng and encloſyng the other diuers ſqua=
res, as in this exāple H.C.E.N. is one ſquare lykeiamme, and
L.M.G.C. is an other, whiche bothe are made aboute one
bias line, that is N.M, than K.L.H.C. and C.E.F.G. are .ij.
fyll ſquares, for they doo fyll vp the ſydes of the .ij. fyrſte
ſquare lykeiammes, in ſuche ſorte, that of all them foure is
made one greate generall ſquare K.M.F.N.

Fyll ſqua=
res.

ἀνάπλη
φώμάτά

Nowe to the ſentence of the theoreme, I ſay, that the .ij.
fill ſquares. H.K.L.C. and C.E.F.G. are both equall togither,
(as it ſhall bee declared in the booke of proofes) bicauſe they
are the fill ſquares of two lykeiammes made aboute one bias
line, as the example ſheweth. Conferre the twelfthe con=
cluſion with this theoreme.

The

GEOMETRICALL.

The xxxiii. Theoreme.

In all right anguled triangles, the square of that side whiche lieth againſt the right angle, is equall to the .ij. ſquares of both the other ſides.

Example.

A.B.C. *is a triangle, hauing a ryght angle in* B. *Wherfore it foloweth, that the ſquare of* A.C, *(whiche is the ſide that lyeth agaynſt the right angle) ſhall be as muche as the two ſquares of* A.B. *and* B.C. *which are the other .ij. ſides. By the ſqnare of any lyne, you muſte vnderſtande a fi= gure made iuſte ſquare, ha= uyng all his iiij. ſydes equall*

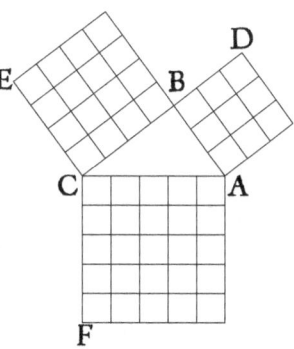

to that line, whereof it is the ſquare, ſo is A.C.F, *the ſquare of* A.C. *Lykewais* A.B.D. *is the ſquare of* A.B. *And* B.C.E. *is the ſquare of* B.C. *Now by the numbre of the diuiſions in eche of theſe ſquares, may you perceaue not onely what the ſquare of any line is called, but alſo that the theoreme is true, and expreſſed playnly bothe by lines and numbre. For as you ſee, the greatter ſquare (that is* A.C.F.) *hath fiue diuiſions on eche ſyde, all equall togyther, and thoſe in the whole ſquare are twenty and fiue. Nowe in the leſt ſquare, whiche is* A.B.D. *there are but .iij. of thoſe diuiſions in one ſyde, and that yeldeth nyne in the whole. So lykeways you ſee in the meane ſquare* A.C.E. *in euery ſyde .iiij. partes, whiche in the whole amount vnto ſixtene. Nowe adde togyther all the partes of the two leſſer ſquares, that is to ſaye, ſixtene and nyne, and you perceyue that they make twnety and fiue, why= che is an equall numbre to the ſumme of the greatter ſquare.*

By

THEOREMES

By this theoreme you may vnderſtand a redy way to know the ſyde of any ryght anguled triangle that is vnknowen, ſo that you knowe the lengthe of any two ſydes of it. For by tournynge the two ſydes certayne into theyr ſquares, and ſo addynge them togyther, other ſubtraɛynge the one from the other (accordyng as in the vſe of theſe theorems I haue ſette foorthe) and then fyndynge the roote of the ſquare that re= mayneth, which roote (I meane the ſyde of the ſquare) is the iuſte length of the vnknowen ſyde, whyche is ſought for. But this appertaineth to the thyrde booke, and therefore I wyll ſpeake no more of it at this tyme.

The xxxiiii. Theoreme.

If ſo be it, that in any triangle, the ſquare of the one ſyde be equall to the .ij. ſquares of the other ij. ſides, than muſt nedes that corner be a right corner, which is conteined betwene thoſe two leſſer ſydes.

Example.

As in the figure of the laſte Theoreme, bicauſe A.C, made in ſquare, is as much as the ſquare of A.B, and alſo as the ſquare of B.C. ioyned bothe togyther, therefore the angle that is in= cloſed betwene thoſe .ij. leſſer lynes, A.B. and B.C. (that is to ſay) the angle B. whiche lieth againſt the line A.C, muſt ne= des be a ryght angle. This theoreme dothe ſo depende of the truthe of the laſte, that whan you perceaue the truthe of the one, you can not iuſtly doubt of the others truthe, for they conteine one ſentence, contrary waies pronounced.

The .xxxv. theoreme.

If there be ſet forth .ij. right lines, and one of them parted into ſundry partes, how many or few

GEOMETRICALL.

*or few ſo euer they be, the ſquare that is made
of thoſe ij. right lines propoſed, is equal to all
the ſquares, that are made of the vndiuided
line, and euery parte of the diuided line.*

Example.

```
    C     D
    ├─────┤

A       E      F   B
├───────┼──────┼───┤

G      M      N  H
┌──────┬──────┬───┐
│      │      │   │
│      │      │   │
K      O      P  L
```

The ij. lines propoſed ar A.
B. *and* C.D, *and the lyne* A.B.
is deuided into thre partes by
E. *and* F. *Now ſaith this theo=
reme, that the ſquare that is
made of thoſe two whole li=
nes* A.B. *and* C.D, *ſo that the
line* A.B. *ſtādeth for the lēgth
of the ſquare, and the other
line* C.D. *for the bredth of the ſame. That ſquare (I ſay) wil be
equall to all the ſquares that be made, of the vndiueded lyne
(which is* C.D.*) and euery portion of the diuided line. And to
declare that particularly, Fyrſt I make an other line* G.K, *e=
quall to the line* C.D, *and the line* G.H. *to be equal to the line*
A.B, *and to bee diuided into iij. like partes, ſo that* G.M. *is e=
quall to* A.E, *and* M.N. *equal to* E.F, *and then muſt* N.H.
nedes remaine equall to F.B. *Then of thoſe ij. lines* G.K, *vn=
deuided, and* G.H. *which is deuided, I make a ſquare, that is*
G.H.K.L, *In which ſquare if I drawe croſſe lines frome one
ſide to the other, according to the diuiſions of the line* G.H,
*then will it appear plaine, that the theoreme doth affirme. For
the firſt ſquare* G.M.O.K, *muſt needes be equal to the ſquare
of the line* C.D, *and the firſt portiō of the diuided line, which
is* A.E, *for bicauſe their ſides are equall. And ſo the ſeconde*

f.i.   *ſquare*

THEOREMES

square that is M.N.P.O, *shall be equall to the square of* C.D, *and the second part of* A.B, *that is* E.F. *Also the third square which is* N.H,L.P, *must of necessitee be equal to the square of* C.D, *and* F.B, *bicause those lines be so coupeled that euery couple are equall in the seuerall figures. And so shal you not only in this example, but in all other finde it true, that if one line be deuided into sondry partes, and another line whole and vndiuided, matched with him in a square, that square which is made of these two whole lines, is as muche iuste and equally, as all the seuerall squares, whiche bee made of the whole line vndiuided, and euery part seuerally of the diui= ded line.*

The xxxvi. Theoreme.

If a right line be parted into ij. partes, as chaunce may happe, the square that is made of that whole line, is equall to bothe the squares that are made of the same line, and the twoo partes of it seuerally.

Example.

The line propouned beyng A.B. *and deuided, as chaunce hap= peneth, in* C. *into ij. vnequall partes, I say that the square made of the hole line* A.B, *is equal to the two squares made of the same line with the twoo partes of itselfe, as with* A.C, *and with* C.B, *for the square* D,E.F.G. *is equal to the two other partial squa res of* D.H.K.G *and* H.E.F.K *but that the greater square is equall to the square of the whole line* A.B, *and the partial*

GEOMETRICALL.

partiall squares equall to the squares of the second partes of the same line ioyned with the whole line, your eye may iudg without muche declaracion, so that I shall not neede to make more exposition therof, but that you may examine it, as you did in the laste Theoreme.

The xxxvii Theoreme.

If a right line be deuided by chaunce, as it maye happen, the square that is made of the whole line, and one of the partes of it which soeuer it be, shal be equall to that square that is made of the ij. partes ioyned togither, and to another square made of that part, which was before ioyned with the whole line.

Example.

The line A.B. is deuided in C. into twoo partes, though not e= qually, of which two partes for an example I take the first, that is A.C, and of it I make one side of a square, as for example D.G.

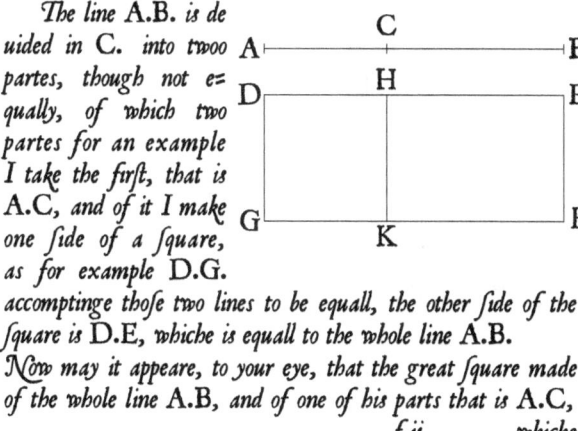

accomptinge those two lines to be equall, the other side of the square is D.E, whiche is equall to the whole line A.B.

Now may it appeare, to your eye, that the great square made of the whole line A.B, and of one of his parts that is A.C,

whiche

THEOREMES

*(which is equall with D,G.) is equal to two partiall squares,
wherof the one is made of the saide greatter portion A.C, in
as muche as not only D.G, beynge one of his sides, but also D.
H. beinge the other side, are eche of them equall to A.C. The
second square is H.E.F.K, in which the one side H.E, is equal
to C.B, being the lesser parte of the line, A.B, and E.F. is e=
quall to A.C. which is the greater parte of the same line. So
that those two squares D.H.K.G, and H.E.F.K, bee bothe of
them no more then the greate square D.E.F.G, accordinge to
the wordes of the Theoreme afore saide.*

The xxxviii. Theoreme.

*If a righte line be deuided by chaunce, into
partes, the square that is made of that whole
line, is equall to both the squares that ar made
of eche parte of the line, and moreouer to two
squares made of the one portion of the diuided
line ioyned with the other in square.*

Example.

*Lette the diuided line bee A.B,
and parted in C, into twoo partes:
Nowe saithe the Theoreme, that
the square of the whole lyne A.B,
is as mouche iuste as the square
of A.C, and the square of C.B, eche
by it selfe, and more ouer by as
muche twife, as A.C. and C.B.*
                                    ioyned

GEOMETRICALL.

ioyned in one square will make. For as you se, the great square
D.E.F.G, *conteyneth in hym foure lesser squares, of whiche
the first and the greatest is* N.M.F.K, *and is equall to the
square of the lyne* A.C. *The second square is the left of them
all, that is* D.H.L.N, *and it is equall to the square of the
line* C.B. *Then are there two other longe squares both of one
bygnes, that is* H.E.N.M. *and* L.N.G.K, *eche of them both
hauyng .ij. sides equall to* A.C, *the longer parte of the diui=
ded line, and there are other two sides equall to* C.B., *beeyng the
shorter parte of the said line* A.B.

So as that greatest square, beeyng made of the hole lyne A.
B, *equal to the ij. squares of eche of his partes seuerally, and
more by as muche iust as .ij. longe squares, made of the lon=
ger portion of the diuided lyne ioyned in square with the
shorter parte of the same diuided line, as the theoreme wold.
And as here I haue put an example of a lyne diuided into .ij.
partes, so the theoreme is true of all diuided lines of what
number so euer the partes be, foure, fyue, or syxe. etc.*
*This theoreme hath great vse, not only in geometrie, but also
in arithmetike, as herafer I will declare in conuenient place.*

The xxxix. theoreme.

*If a right line be deuided into two equall par=
tes, and one of thefe .ij. partes diuided agayn
into two other partes, as happeneth the longe
square that is made of the thyrd or later part
of that diuided line, with the refidue of the
fame line, and the square of the mydlemofte
parte, are bothe togither equall to the square
of halfe the firfte line.*

f.iij.   Example.

THEOREMES

Example.

The line A.B. *is diuided into*
ij. equal partes in C, *and*
that parte C.B. *is diui=*
ded agayne as hapneth
in D. *Wherfore faith the*
Theorem that the long
fquare made of D.B.
and A.D, *with the fqua*
re of C.D. *(which is the*

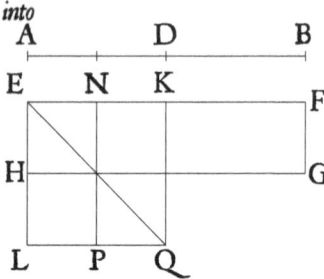

mydle portion) fhall bothe be equall to the fquare of half the
lyne A.B, *that is to faye, to the fquare of* A.C, *or els of* C.
D, *which make all one. The long fquare* F.G.N.O. *whiche is*
the longe fquare that the theoreme fpeaketh of, is made of .ij.
long fquares, wherof the fyrft is F.G.M.K, *and the fecond*
is K.N.O.M. *The fquare of the myddle portion is* L.M.
O.P. *And the fquare of the halfe of the fyrfte lyne is* E.K.
Q.L. *Nowe by the theoreme, that longe fquare* F.G.M.
O, *with the iufte fquare* L.M.O.P, *mufte bee equall to the*
greate fquare E.K.Q.L, *whyche thynge bycaufe it feemeth*
fomewhat difficult to vnderftande, althoughe I intende not
here to make demonftrations of the Theoremes, bycaufe it
is appoynted to be done in the newe edition of Euclide, yet I
wyll fhew you brefely how the equalitee of the partes doth
ftande. And fyrft I fay, that where the comparyfon of equa=
litee is made betweene the greate fquare (whiche is made of
halfe the line A.B.) *and two other, where of the fyrft is the*
longe fquare F.G.N.O, *and the feconde is the full fquare* L.
M.O.P, *which is one portion of the great fquare allredye,*
and fo is that longe fquare K.N.M.O, *beynge a parcell alfo*
of the longe fquare F.G.M.O. *Wherfore as thofe two par=*
tes are common to bothe partes compared in equalitee, and
therfore beynge bothe abated from eche parte, if the refte of
bothe the other partes bee equall, than were thofe whole par
tes equall before : Nowe the refte of the great fquare, thofe

two

GEOMETRICALL.

two leſſer ſquares beyng taken away) is that longe ſquare E.
N.P.Q, *whyche is equall to the long ſquare* F.G.K.M, *be=*
yng the reſt of the other parte. And that they two be equall,
theyr ſydes doo declare. For the longeſt lynes that is F.K, *and*
E.Q. *are equall, and ſo are the ſhorter lynes,* F.G, *and* E.N,
and ſo appereth the truthe of the Theoreme.

The .xl. theoreme.

If a right line be diuided into .ij. euen par=
tes, and an other right line annexed to one ende
of that line, ſo that it make one righte line
with the firſte. The longe ſquare that is made
of this whole line ſo augmented, and the por=
tion that is added, with the ſquare of halfe the
right line, ſhall be equall to the ſquare of that
line, whiche is compounded of halfe the firſte
line, and the parte newly added.

Example.

The fyrſt lyne propouned is
A.B, *and it is diuided into*
ij. equall partes in C, *and an*
other ryght lyne, I meane
B.D, *annexed to one ende*
of the fyrſte lyne.
Nowe ſay I, *that the long*
ſquare A.D.M.K, *is made*
of the whole lyne ſo aug=
meted, that is A.D, *and the portiō annexed,* ẏ *is* D.M, *for* D.M
is equall to B.D, *wherfore* ẏ *long ſquare* A.D.M.K, *with the*
ſquare

THEOREMES

square of halfe the firſt line, that is E.G.H.L, *is equall to the great ſquare* E.F.D.C. *whiche ſquare is made of the line* C. D. *that is to ſaie, of a line compounded of halfe the firſt line, beyng* C.B, *and the portion annexed, that is* B.D. *And it is eaſyly perceaued, if you conſyder that the longe ſquare* A.C. L.K. *(whiche onely is lefte out of the great ſquare) hath a nother longe ſquare equall to hym, and to ſupply his ſteede in the great ſquare, and that is* G.F.M.H. *For their ſydes be of lyke lines in length.*

The xli. Theoreme.

If a right line bee diuided by chaunce, the ſquare of the ſame whole line, and the ſquare of one of his partes are iuſte equall to the lōg ſquare of the whole line, and the ſayde parte twiſe taken, and more ouer to the ſquare of the other parte of the ſayd line.

Example.

A.B. *is the line diuided in* C. *And* D.E.F.G, *is the ſquare of the whole line,* D.H.K.M. *is the ſquare of the leſſer portion (whyche I take for an example) and therfore muſt bee twiſe reckened. Nowe I ſaye that thoſe* ij. *ſquares are equall to two longe ſquares of the whole line* A. B, *and his ſayd portion* A.C, *and alſo to the ſquare of the other portion of the ſayd firſt line, whiche porti= on is* C.B, *and his ſquare* K.N.F.L. In this theoreme there is no difficultie, if you cōſyder that the litle ſquare D.H.K.M. is .iiij. tymes reckened, that is to ſay, fyrſt of all as a parte of the greateſt ſquare, whiche is D.E.F.G. Secondly he is rekned by

GEOMETRICALL.

by himfelfe. Thirdely he is accompted as parcell of the long fquare D.E.N.M, *And fourthly he is taken as a part of the o= ther long fquare* D.H.L.G, *fo that in as muche as he is twife reckened in one part of the comparifō of equalitee, and twife alfo in the fecond parte, there can rife none occafion of error or doubtfulnes therby.*

The xlii. Theoreme.

If a right line be deuided as chance happe= neth the iiij. long fquares, that may be made of that whole line and one of his partes with the fquare of the other part, fhall be equall to the fquare that is made of the whole line and the faide firft portion ioyned to him in lengthe as one whole line.

Example.

The firfte line is A.B, *and is deuided by* C. *into two vn= equall partes as happeneth. the longfquare of yt and his leffer portion* A.C, *is foure times drawen, the firft is* E. G.M.K, *the feconde is* K. M.Q.O, *the third is* H.K.R. S, *and the fourthe is* K.L.S. T. *And where as it appea= reth that one of the little fquares* (I meane K.L.P O) *is reckened twife, ones as par cell of the fecond longfquare and agayne as parte of the*

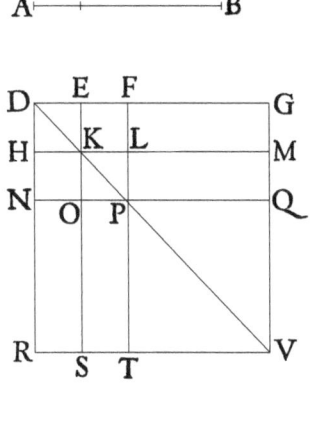

g.i.   thirde

THEOREMES

thirde longſquare, to auoide ambiguite, you may place one
in ſteede of it, an other ſquare of equalitee, with it. that is
to ſaye, D.E.K.H, which was at no tyme accomptyng as par=
cell of any one of them, and then haue you iiij. long ſquares di
ſtinctly made of the whole line A.B, and his leſſer portion
A.C. And within them is there a greate full ſquare P.Q.T.V.
whiche is the iuſt ſquare of B.C, beynge the greatter portion
of the line A.B. And that thoſe fiue ſquares doo make iuſte as
muche as the whole ſquare of that longer line D.G, (whiche
is as longe as A.B, and A.C. ioyned togither) it may be iudged
eaſyly by the eye, ſith that one greate ſquare doth comprehed
in it all the other fiue ſquares, that is to ſay, foure longſquares
(as is before mencioned) and one full ſquare which is the in=
tent of the Theoreme.

The xliii. Theoreme.

If a right line be deuided into ij. equal par=
tes firſt, and one of thoſe parts again iuto o=
ther ij. parts, as chaūce hapeneth, the ſquare
that is made of the laſt part of the line ſo di=
uided, and the ſquare of the reſidue of that
whole line, are double to the ſquare of halfe
that line, and to the ſquare of the middle por=
tion of the ſame line.

Example.

The line to be deuided is A.B, and is parted in C. into two
equall partes, and then C.B, is deuided againe into two par
tes in D, ſo that the meaninge of the Theoreme, is that the
ſquare

GEOMETRICALL.

square of D.B. *which is the latter parte of the line, and the*

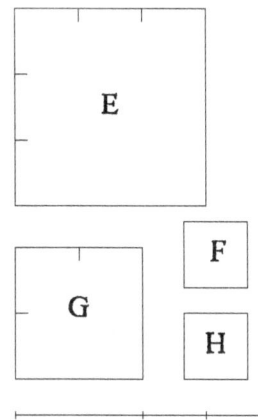

square of A.D, *which is the residue of the whole line. Those two squares, I say, ar double to the square of the one halfe of the line, and to the square of* C.D, *which is the middle portion of those thre diuisions. Which thing that you maye more easilye perceaue, I haue drawen foure squares, whereof the greatest being marked with* E. *is the square of* A.D. *The next, which is marked with* G, *is the square of halfe the line, that is, of* A.C, *And the other two little squares marked with* F. *and* H, *be both of one big= nes, by reason that I did diuide* C.B. *into two equall partes, so that you may take the square* F, *for the square of* D.B, *and the square* H, *for the square of* C.D. *Now I thinke you doubt not, but that the square* E. *and the square* F, *ar double so much as the square* G. *and the square* H, *which thing the eayser is to be vnderstande, bicause that the greate square hath in his side iiij. quarters of the firste line, whiche multiplied by it selfe maketh nyne quarters, and the square* F. *containeth but one quarter, so that bothe doo make tenne quarters.*

Then G. *contayneth iiij. quarters, seynge his syde containeth twoo, and* H. *containeth but one quarter, whiche both make but fiue quarters, and that is but halfe of tenne.*

Whereby you may easilye coniecture, that the meanynge of the the= oreme is verified in the figures of this ex= ample.

THEOREMES

The xliiii. Theoreme.

If a right line be deuided into ij. partes e=
qually, and an other portion of a righte lyne
annexed to that firſte line, the ſquare of this
whole line ſo compounded, and the ſquare of
the portion that is annexed, ar double as much
as the ſquare of the halfe of the firſte line,
and the ſquare of the other halfe ioyned in
one with the annexed portion, as one whole
line.

Example.

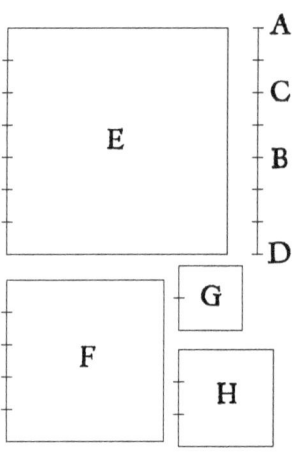

A

C

B

D

G

E

F

H

The line is **A.B,** *and is di*
uided firſte into twoo e=
qual partes in **C,** *and thē*
is there annexed to it an
other portion whiche is
B.D. *Now ſaith the The*
oreme, that the ſquare
of **A.D,** *and the ſquare*
of **B.D,** *ar double to the*
ſquare of **A.C,** *and to the*
ſquare of **C.D.** *The line*
A.B. *cōtaining four par*
tes, then muſt needes his
halfe containe ij. partes
of ſuch partes I ſuppoſe
B.D. *(which is the ānex*
ed line) to containe thre, ſo ſhall the hole line cōprehend vij.
partes, and his ſquare xlix. partes, wherunto if you ad ẏ ſquare
of

GEOMETRICALL.

*of the annexed lyne, whiche maketh nyne, than thofe bothe
doo yelde, lviij. whyche muft be double to the fquare of the
halfe lyne with the annexed portion. The halfe lyne by itfelfe
conteyneth but .ij. partes, and therfore his fquare dooth make
foure. The halfe lyne with the annexed portion conteyneth
fiue, and the fquare of it is .xxv. now put foure to .xxv, and
it maketh iuft .xxix, the euen halfe of fifty and eight, wher
by appereth the truthe of the theoreme.*

The .xlv. theoreme.

*In all triangles that haue a blunt angle, the
fquare of the fide that lieth againft the blunt
angle, is greater than the two fquares of the
other twoo fydes, by twife as muche as is
comprehended of the other of thofe .ij. fides (in=
clofyng the blunt corner) and that portion of
the fame line, beyng drawen foorth in lengthe,
which lieth betwene the faid blunt corner and
a perpendicular line lightyng on it, and dra=
wen from one of the fharpe angles of the fore=
fayd triangle.*

Example.

*For the declaration of this theoreme, and the next alfo, whofe
vfe are wonderfull in the practife of Geometrie, and in mea
furyng efpecially, it fhall be nedefull to declare that euery tri
angle that hath no ryght angle, as thofe be whyche are called
(as in the boke of practife is declared) fharp cornered trian=
gles, and blunt conered triangles, yet may they be brought to
haue a ryght angle, eyther by partyng them into two leffer tri*

g.iij.   angles

THEOREMES

angles, or els by addyng an other triangle vnto them, whiche
may be a great helpe for the ayde of meafuryng, as more large
ly fhall be fette foorthe in the boke of meafuryng. But for this
prefent place, this forme wyll I vfe, (which Theon alfo v=
feth) to adde one triangle vnto an other, to bryng the blunt
cornered triangle into a ryght angled triangle, whereby the
proportion of the fquares of the fides in fuche a blunte cor=
nered triangle may the better bee knowen.

Fyrft ther=
fore I fette
foorth the tri
angle A.B.C,
whofe cor=
ner by C. is a
blunt corner
as you maye
well iudge,
than to make
an other tri=
angle of yt
with a ryght
angle, I muft
drawe forth
the fide B.C.
vnto D, and
frō the fharp
corner by A.
I brynge a
plumbe lyne
or perpēdi=
cular on D. And fo is there nowe a newe triangle A.B.D.
whofe angle by D. is a right angle. Nowe accordyng to the
meanyng of the Theoreme, I faie, that in the firft triangle A.
B.C, becaufe it hath a blunt corner at C, the fquare of the
line A.B. whiche lieth againft the faid blunte corner, is more
                                                          then

GEOMETRICALL.

then the square of the line A.C, *and also of the lyne* B.C. (*whiche inclose the blunte corner*) *by as muche as will amount twise of the line* B.C, *and that portion* D.C. *whiche lieth be= twene the blunt angle by* C, *and the perpendicular line* A.D.

The square of the line A.B, *is the great square marked with* E. *The square of* A.C, *is the meane square marked with* F. *The square of* B.C, *is the least square marked with* G. *And the long square marked with* K, *is sette in steede of two squares made of* B.C, *and* C.D. *For as the shorter side is the iuste lengthe of* C.D, *so the other longer side is iust twise so longe as* B.C, *Wherfore I saie now accordyng to the Theo= reme, that the greatte square* E, *is more then the other two squares* F. *and* G, *by the quantitee of the longe square* K, *wherof I reserue the profe to a more conuenient place, where I will also teache the reason howe to fynde the lengthe of all suche perpendicular lynes, and also of the line that is drawen betweene the blunte angle and the perpendicular line, with sundrie other very pleasant conclusions.*

The .xlvi. Theoreme.

In sharpe cornered triangles, the square of anie side that lieth against a sharpe corner, is lesser then the two squares of the other two sides, by as muche as is comprised twise in the long square of that side, on whiche the perpen= dicular line falleth, and the portion of that same line, liyng betweene the perpendicular, and the foresaid sharpe corner.

Example.

First

THEOREMES

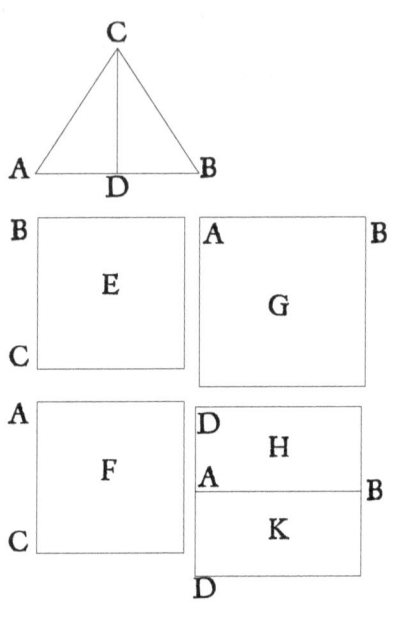

Fyrſt I ſette foorth the tri= angle A.B.C, and in yt I draw a plūbe line from the angle C. vnto the line A.B, and it lighteth in D. Nowe by the theore= me the ſquare of B.C. is not ſo muche as the ſquare of the other two ſydes, that of B.A. and of A.C. by as mu che as is twiſe conteyned in the lōg ſquare made of A.B, and A.D, A.B. beyng the line or ſyde on which the perpendicular line falleth, and A.D. beeyng that portion of the ſame line whiche doth lye betwene the perpendicular line, and the ſayd ſharpe angle limitted, whiche angle is by A.

For declaration of the figures, the ſquare marked with E. is the ſquare of B.C, whiche is the ſyde that lieth agaynſt the ſharpe angle, the ſquare marked with G. is the ſquare of A. B, and the ſquare marked with F. is the ſquare of A.C, and the two longe ſquares marked with H. K, are made of the hole line A.B, and one of his portions A.D. And truthe it is that the ſquare E. is leſſer than the other two ſquares G. and F. by the quantitee of thoſe two long ſquares H. and K. Wher by you may conſyder agayn, an other proportion of equalitee,
                                                                    that

GEOMETRICALL.

that is to faye, that the fquare E. *with the twoo long fquares*
H.K, *are iufte equall to the other two fquares* G. *and* F. *And
fo maye you make, as it were an other theoreme.* That in al
fharpe cornered triangles, where a perpendicular
line is drawen frome one angle to the fide that ly=
eth againfte it, the fquare of anye one fide, with the
ii. longe fquares made of that hole line, whereon the
perpendicular line doth lighte, and of that portion
of it, which ioyneth to that fide, whole fquare is all
ready taken, thofe thre figures, I fay, are equall to the
ii. fquares, of the other ii. fides of the triangle. *In
whiche you mufte vnderftand, that the fide on which the per
pendiculare falleth, is thrife vfed, yet is his fquare but ones
mentioned, for twife he is taken for one fide of the two long
fquares. And as I haue thus made as it were an other theo=
reme out of this fourty and fixe theoreme, fo mighte I out of
it, and the other that goeth nexte before, make as manny as
woulde fuffice for a whole booke fo that when they fhall bee
applyed to practife, and confequently to expreffe their bene=
fite, no manne that hathe not well wayde their wonderfull
commoditee, woulde credite the pofibilitie of their wonder=
full vfe, and large ayde in knowledge. But all this wyll I re=
mitte to a place conuenient.*

The xlvii. Theoreme.

*If ij. points be marked in the circumferēce
of a circle, and a right line drawen frome the
one to the other, that line muft needes fal with
in the circle.*

Example.

The circle is A.B.C.D, *the ij. pointes are* A. B, *the righte*
b.i.   *line*

THEOREMES

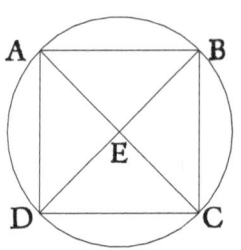

line that is drawenne frome the one to the other, is the line A.B, which as you see, must needes lyghte within the circle. So if you putte the pointes to be A.D, or D.C, or A.C, other B.C, or B.D, in any of these ca ses you see, that the line that is dra= wen from the one pricke to the other dothe euermore run within the edge of the circle, els canne it be no right line. How be it, that a croked line, especially being more cro ked then the portion of the circumference, maye bee drawen from pointe to pointe withoute the circle. But the theoreme speaketh only of right lines, and not of croked lines.

The xlviii. Theoreme.

If a righte line passinge by the centre of a circle, doo crosse an other right line within the same circle, passinge beside the centre, if be deuide the saide line into twoo equal partes, then doo they make all their angles righte. And contrarie waies, if they make all their angles righte, then doth the longer line cutte the shorter in twoo partes.

Example.

The circle is A.B.C.D, the line that passeth by the centre, is A.E.C, the line that goeth beside the centre is D.B. Nowe say

GEOMETRICALL.

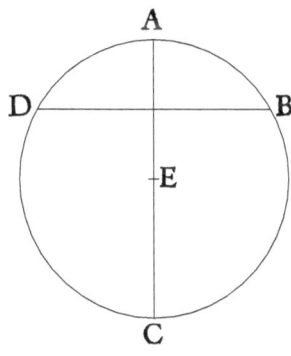

ſaye I, that the line A.E.
C, *dothe cutte that other
line* D.B. *into twoo iuſte
partes, and therefore all
their four angles ar righte
angles. And contrarye
wayes, bicauſe all their
angles are righte angles,
therfore it muſte be true,
that the greater cutteth
the leſſer into two equal
partes, acordinge as the*

Theoreme would.

The xlix. Theoreme.

*If twoo right lines drawen in a circle doo
croſſe one an other, and doo not paſſe by the
centre, euery of them dothe not deuide the o=
ther into two equall partions.*

Example.

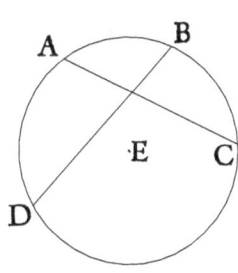

The circle is A.B.C.D, *and the cen
tre is* E, *the one line* A.C, *and the
other is* B.D, *which two lines croſſe
one an other, but yet they go not by
the centre, wherefore accordinge
to the woordes of the theoreme,
eche of theim doth cutte the other
into equall portions. For as you may
eaſily iudge,* A.C. *hath one portiõ lõ
ger and an other ſhorter, and ſo like
wiſe* B.D. *Howbeit, it is not ſo to be
vnderſtãd, but one of them may be diuided into ij. euẽ parts,*

h.ij.   but

THEOREMES

*but bothe to bee cutte equally in the middle, is not poſſible,
onles both paſſe through the cētre, therfore much rather whē
bothe go beſide the centre, it can not be that eche of theym
ſhoulde be iuſtely parted into ij. euen partes.*

The L. Theoreme.

*If two circles croſſe and cut one an other,
then haue not they both one centre.*

Example.

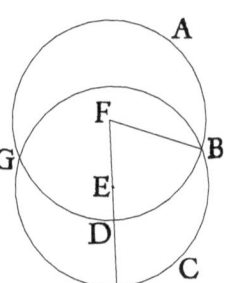

*This theoreme ſeemeth of it ſelfe
ſo manifeſt, that it neadeth nother
demonſtration nother declaraciō.
Yet for the plaine vnderſtanding
of it, I haue ſette forthe a figure
here, where ij. circles be drawē,
ſo that one of them doth croſſe the
other (as you ſee) in the pointes
B. and G, and their centres appear
at the firſte ſighte to bee diuers. For the centre of the one is F,
and the centre of the other is E, which diffre as farre a ſondre,
as the edges of the circles, where they bee moſte diſtaunte in
ſonder.*

The Li. Theoreme.

*If two circles be ſo drawen, tht one of them
do touche the other, then haue they not one
centre.*

Exam

GEOMETRICALL.

Example.

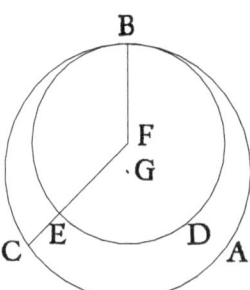

There are two circles made, as you see, the one is A.B.C, and hath his centre by G, the other is B.D.E, and his centre is by F, so that it is eaſy enough to perceiue that their centres doe dyffer as muche a ſonder, as the halfe dia=meter of the greater circle is lō=ger then the half diameter of the leſſer circle. And ſo muſt it nee=des be thought and ſaid of all o=ther circles in lyke kinde.

The .lii. theoreme.

If a certaine pointe be aſſigned in the dia=meter of a circle, diſtant from the centre of the ſaid circle, and from that pointe diuerſe lynes drawen to the edge and circumference of the ſame circle, the longeſt line is that whiche paſ=ſeth by the centre, and the ſhorteſt is the reſi=dew of the ſame line. And of al the other lines that is euer the greateſt, that is nigheſt to the line, which paſſeth by the centre. And cōtra ry waies, that is ſhorteſt, that is fartheſt from it. And amongeſt thē all there can be but one=ly .ij. equall together, and they muſt nedes be ſo placed, that the ſhorteſt line ſhall be in the iuſt middle betwixte them.

h.iij.   Ex=

THEOREMES

Example.

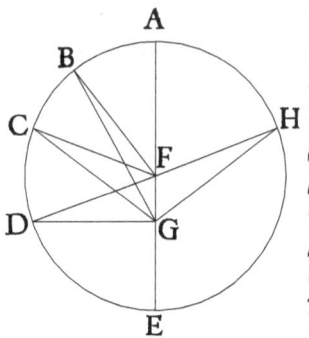

The circle is A.B.C.D.E.H, *and his centre is* F, *the diame= ter is* A.E, *in whiche diameter I haue taken a certain point di= ftaunt from the centre, and that pointe is* G, *from whiche I haue drawen .iiij. lines to the circum= ference, befide the two partes of the diameter, whiche maketh vp vi. lynes in all. Nowe for the diuerfitee in quantitie of thefe lynes, I faie accordyng to the Theoreme, that the line whiche goeth by the centre is the longeft line, that is to faie,* A.G, *and the refidewe of the fame diameter beeyng* G.E, *is the fhorteft lyne. And of all the other that lyne is longeft, that is neereft vnto that parte of the diameter whiche gooeth by the centre, and that is fhorteft, that is fartheft diftant from it, wherefore I faie, that* G.B, *is longer then* G.C, *and therfore muche more longer then* G.D, *fith* G.C, *alfo is longer then* G.D, *and by this maie you foone perceiue, that it is not poffible to drawe .ij. lynes on any one fide of the diameter, which might be equall in lengthe together, but on the one fide of the diameter maie you eafylie make one lyne equall to an other, on the other fide of the fame diameter, as you fee in this example* G.H, *to bee equall to* G.E, *betweene whiche the lyne* G.E, *(as the fhorteft in all the circle) doothe ftande euen diftaunte from eche of them, and that is the precife knowledge of their equalitee, if they be equally diftaunt from one halfe of the diameter. Where as contrary waies if the one be neerer to any one halfe of the diameter then the other is, it is not poffible that they two may be equall in lengthe, namely if they dooe ende bothe in the circumference of the*
circle,

GEOMETRICALL.

circle, and be bothe drawen from one poynte in the dia=
meter, fo that the faide poynte be (as the Theoreme doeth
fuppofe) fomewhat diftaunt from the centre of the faid cir=
cle. For if they be drawen from the centre, then mu/t they
of neceffitee be all equall, howe many fo euer they bee,
as the definition of a circle dooeth importe, withoute any
regarde how neere fo euer they be to the diameter, or how
diftante from it. And here is to be noted, that in this The=
oreme, by neereneffe and diftaunce is vnderftand the nere=
neffe and diftaunce of the extreeme partes of thofe lynes
where they touche the circumference. For at the other end
they do all meete and touche.

The .liii. Theoreme.

If a pointe bee marked without a circle,
and from it diuerfe lines drawen croffe the
circle, to the circumference on the other
fide, fo that one of them paffe by the centre,
then that line whiche paffeth by the centre
fhall be the longeft of all them that croffe the
circle. And of thother lines thofe are longeft,
that be nexte vnto it that paffeth by the centre.
And thofe ar fhorteft, that be fartheft diftant
from it. But among thofe partes of thofe lines,
whiche ende in the outewarde circumference,
that is moft fhorteft, whiche is parte of the line
that paffeth by the centre, and amongefte the

other

THEOREMES

othere eche, of thē, the nerer they are vnto it,
the ſhorter they are, and the farther from it,
the longer they be. And amongeſt them all
there can not be more then .ij. of any one lēgth,
and they two muſte be on the two contrarie ſi=
des of the ſhorteſt line.

Example.

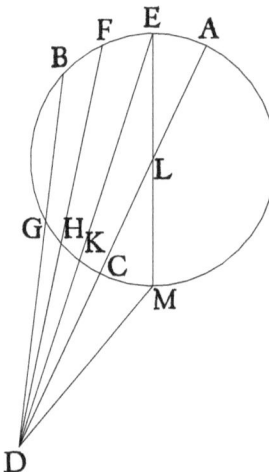

Take the circle to be A.B.C,
and the point aſſigned without it
to be D. *Now ſay I, that if there*
be drawen ſundrie lines from D,
and croſſe the circle, endyng in
the circumference on the cōtrary
ſide, as here you ſee, D.A, D.E,
D.F, *and* D.B, *then of all theſe*
lines the longeſt muſt needes be
D.A, *which goeth by the centre*
of the circle, and the nexte vnto
it, that is D.E, *is the longeſt a=*
mongeſt the reſt. And contrarie
waies, D.B, *is the ſhorteſte, be=*
cauſe it is fartheſt diſtaunt from
D.A. *And ſo maie you iudge of* D.F, *becauſe it is nerer vnto*
D.A, *then is* D.B, *therefore is it longer then* D.B. *And like=*
waies becauſe it is farther of from D.A, *then is* D.E, *therfore*
is it ſhorter then D.E. *Now for thoſe partes of the lines whi=*
che bee withoute the circle (as you ſee) D.C, *is the ſhorteſt,*
becauſe it is the parte of that line which paſſeth by the centre,
And D.K, *is next to it in diſtance, and therfore alſo in ſhortnes,*
ſo D.G, *is fartheſt from it in diſtance, and therfore is the lon=*
geſt of them. Now D.H, *beyng nerer then* D.G, *is alſo ſhor=*
ter

GEOMETRICALL.

ter then it, and beynge farther of, then D.K, *is longer then it.*
ſo that for this parte of the theoreme (as I thinĸ) you do plain
ly perceaue the truthe thereof, ſo the reſidue hathe no dif=
ficulte. For ſeing that the nearer any line is to D.C, *(which ioy*
neth with the diameter) the ſhorter it is and the farther of
from it, the longer it is. And ſeyng two lynes can not be of liĸe
diſtaunce beinge bothe on one ſide, therefore if they ſhal be
of one lengthe, and conſequently of one diſtaunce, they muſt
needes bee on contrary ſides of the ſaide line D.C. *And ſo ap=*
peareth the meaning of the whole Theoreme.

And of this Theoreme dothe there folowe an other lyĸe,
whiche you maye calle other a theoreme by itſelfe, or elſe
a Corollary vnto this laſte theoreme, I paſſe not ſo muche
for the name. But his ſentence is this : when ſo euer any ly=
nes be drawen frome any pointe, withoute a circle,
whether they croſſe the circle, or eande in the utter
edge of his circumference, thoſe two lines that bee e=
qually diſtaunt from the leaſt line are equal togither,
and contrary waies, if they be equall togither, they
ar alſo equally diſtant from the leaſt line.

For the declaracion of this propoſition, it ſhall
not need to vſe any other example, then that which is brought
for the explication of this laſte theoreme, by whiche you may
without any teachinge eaſyly perceaue both the meanyng and
alſo the truth of this propoſition.

The Liiii. Theoreme.

If a point be ſet forthe in a circle, and frō
that pointe vnto the circumference many li=
nes drawen, of which more then two are equal
togither, then is that point the centre of that
circle.

i.i.   *Example*

THEOREMES

Example.

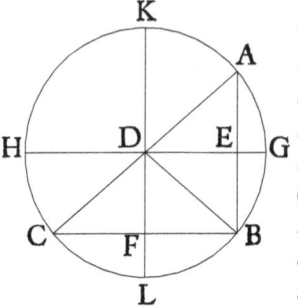

The circle is A.B.C, *and with
in it I haue fette fourth for an
example three prickes, which
are* D.E. *and* F, *and from eue=
ry one of them I haue drawē
(at the leafte) iiij. lines vnto
the circumference of the circle
but frome* D, *I haue drawen
more, yet maye it appear rea=
dily vnto your eye, that of all
the lines whiche be drawen from* E. *and* F, *vnto the circum=
ference, there are but twoo equall, and more can not bee, for*
G.E. *nor* E.H. *hath none other equal to theim, nor canne not
haue any beinge drawen from the fame point* E. *No more can*
L.F, *or* F.K, *haue anye line equall to either of theim, beinge
drawen from the fame pointe* F. *And yet from either of thofe
two pointes are there drawen twoo lines equall togither,
as* A.E, *is equall to* E.B, *and* B.F, *is equall to* F.C, *but there
can no third line be drawen equall to either of thefe two cou=
ples, and that is by reafon that they be drawen from a pointe
diftaunte from the centre of the circle. But from* D *althoughe
there be feuen lines drawen, to the circumference, yet all bee
equall, bicaufe it is the centre of the circle. And therefore if
you drawe neuer fo mannye more from it vnto the circumfe=
rence, all fhall be equal, fo that this is the priuilege (as it were
of the centre) and therfore no other point can haue aboue two
equal lines drawen from it vnto the circumference. And from
all pointes you maye drawe ij. equall lines to the circumfe=
rence of the circle, whether that pointe be within the circle
or without it.*

The lv. Theoreme.

*No circle canne cut an other circle in more
pointes*

GEOMETRICALL.

pointes then two.

Example.

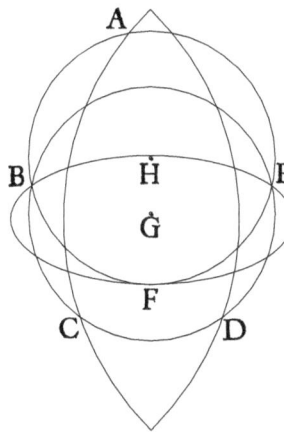

The firſt circle is A.B.F.E, *the ſecond circle is* B.C.D, E, *and they croſſe one an o=ther in* B. *and in* E, *and in no more pointes.* Nother *is it poſſible that they ſhould, but other figures ther be, which maye cutte a circle in foure partes, as you ſe in this exā= ple. Where I haue ſet forthe one tunne forme, and one eye forme, and eche of them cut teth euery of their two cir= cles into foure partes. But as they be irregulare formes, that is to ſaye, ſuche formes as haue no preciſe meaſure nother proportion in their draughte, ſo can there ſcarſely be made any certaine theorem of them. But circles are regulare formes, that is to ſay, ſuch formes as haue in their protraĉture a iuſte and certaine proportion, ſo that certain and determinate truths may be affirmed of them, ſith they ar vniforme and vnchaungable.*

The lvi. Theoreme.

If two circles be ſo drawen, that the one be within the other, and that they touche one an other: If a line bee drawen by bothe their centres, and ſo forthe in lengthe, that line ſhall runne to that pointe, where the circles do touche.

i.ij.   Example

THEOREMES

Example

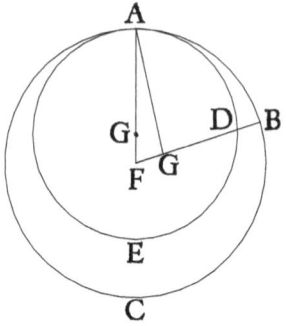

The one circle, which is the greatteſt and vttermoſt is A. B.C, the other circle that is $\overset{e}{y}$ leſſer, and is drawen within the firſte, is A.D.E. The cētre of the greater circle is F, and the centre of the leſſer circle is G, the pointe where they touche is A. And now you may ſee the truthe of the theoreme ſo plainely, that it needeth no farther declaracion. For you maye ſee, that drawinge a line frome F. to G, and ſo forth in lengthe, vntill it come to the circumference, it wyll lighte in the very pointe A, where the circles touch one an other.

The Lvii. Theoreme.

If two circles bee drawen ſo one withoute and other, that their edges doo touche and a right line bee drawenne frome the centre of the one to the centre of the other, that line ſhall paſſe by the place of their touching.

Example.

The firſte circle is A.B.E, and his centre is K, The fecōd circle is D,B.C, and his cētre is H, the point wher they do touch is B. Nowe doo you ſe that the line K.H, whiche is drawen frome

GEOMETRICALL.

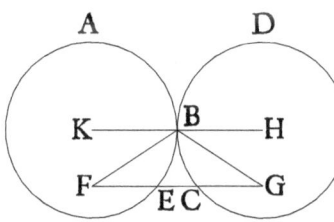

from K, that is cen=
tre of the firſte cir=
cle, vnto H, beyng
centre of the ſecond
circle, doth paſſe (as
it muſt nedes by the
pointe B,) whiche is
the verye poynte
wher they doth tu=
che together.

The .lviii. theoreme.

*One circle can not touche an other in more
pointes then one, whether they touche within
or without.*

Example.

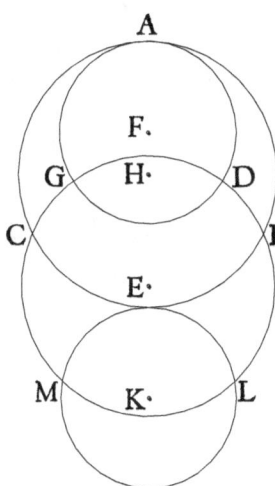

For the declaration of this
Theoreme, I haue drawen
iiij. circles, the firſt is A.B.C,
and his centre H. the ſecond
is A.D.G, and his centre F.
the third is L.M, and his cen
tre K. the .iiij. is D.G.L.M,
and his centre E. Nowe as
you perceiue the ſecond cir=
cle A.D.G, toucheth the firſt
in the inner ſide, in ſo much
as it is drawen within the o=
ther, and yet it toucheth him
but in one point, that is to ſay
in A, ſo lykewaies the third
circle L.M. is drawen without the firſte circle and toucheth

THEOREMES

hym, as you maie fee, but in one place. And now as for the .iiij.
circle, it is drawen to declare the diuerſitie betwene touchyng
and cuttyng, or croſſyng. For one circle maie croſſe and cutte a
great many other circles, yet can he not cutte any one in more
places then two, as the fiue and fiftie Theoreme affirmeth.

The .lix. Theoreme.

In euerie circle thoſe lines are to be counted
equall, whiche are in lyke diſtaunce from the
centre, And contrarie waies they are in lyke
diſtance from the centre, whiche be equall.

Example.

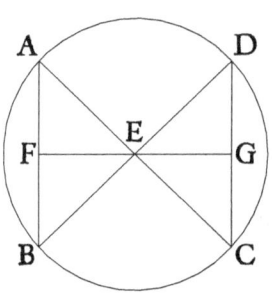

In this figure you ſee firſte
the circle drawen, whiche is
A.B.C.D, *and his centre is* E.
In this circle alſo there are
drawen two lines equally
diſtaunt from the centre, for
the line A.B, *and the line* D.
C, *are iuſte of one diſtaunce*
from the centre, whiche is E,
and therfore are they of one
length. Again thei are of one
lengthe (as ſhall be proued in the boke of profes) and therefore
their diſtaunce from the centre is all one.

The lx. Theoreme.

In euerie circle the longeſt line is the diame=
ter, and of all the other lines, thei are ſtill lon=
$$\qquad\qquad\qquad\qquad\qquad\qquad geſt$$

GEOMETRICALL.

*geſt that be nexte vnto the centre, and they be
the ſhorteſt, that be fartheſt diſtaunt from it.*

Example.

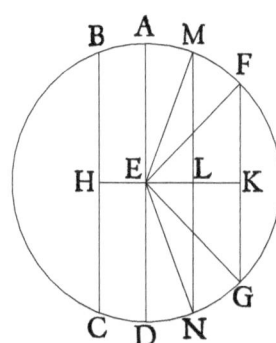

In this circle A.B.C.D, *I
haue drawen firſt the diame=
ter, whiche is* A.D, *whiche
paſſeth (as it muſt) by the cen
tre* E, *Then haue I drawen
ij. other lines as* M.N, *whi=
che is neerer the centre, and*
F.G, *that is farther from the
centre, The fourth line alſo
on the othr ſide of the dia=
meter, that is* B.C, *is neerer
to the centre then the line* F.
G, *for it is of lyke diſtance as
is the line* M.N. *Nowe ſaie I, that* A.D, *beyng the diame=
ter, is the longeſt of all thoſe lynes, and alſo of any other that
maie be drawen within that circle, And the other line* M.N,
is longer then F.G, *becauſe it is nerer to the centre of the cir=
cle then* F.G. *Alſo the line* F.G, *is ſhorter then the line* B.C.
for becauſe it is farther from the centre then is the lyne B.C.
*And thus maie you iudge of al lines drawen in any circle, how
to know the proportion of their length, by the proportion of
their diſtance, and contrary waies, howe to diſcerne the pro=
portion of their diſtance by their lengthes, if you knowe the
proportion of their length. And to ſpeake of it by the waie,
it is a maruaylouſe thyng to conſider, that a man maie knowe
an exacte proportion betwene two thynges, and yet can not
name nor attayne the preciſe quantitee of thoſe two thynges,
As for exaumple, If two ſquares be ſette foorthe, whereof the
one containeth in it fiue ſquare feete, and the other contayneth
fiue and fortie foote, of like ſquare feete, I am not able to tell,
no nor yet anye manne liuyng, what is the preciſe mea=*
                                                        ſure

THEOREMES

*fure of the ſides of any of thoſe .ij. ſquares, and yet I can proue
by vnfallible reaſon, that their ſides be in a triple proportion,
that is to ſaie, that the ſide of the greater ſquare (whiche con=
taineth .xlv. foote) is three tymes ſo long iuſte as the ſide of the
leſſer ſquare, that includeth but fiue foote. But this ſeemeth to
be ſpoken out of ceaſon in this place, therfore I will omitte it
now, referuyng the exacter declaration therof to a more con=
uenient place and time, and will procede with the reſidew of
the Theoremes appointed for this boke.*

The .lxi. Theoreme.

*If a right line be drawen at any end of a di=
ameter in perpendicular forme, and do make a
right angle with the diameter, that right line
ſhall light without the circle, and yet ſo ioint=
ly knitte to it, that it is not poſſible to draw a=
ny other right line betwene that ſaide line and
the circumferēce of the circle. And the angle
that is made in the ſemicircle is greater then
any ſharpe angle that may be made of right li=
nes, but the other angle without, is leſſer then
any that can be made of right lines.*

Example.

*In this circle A.B.C, the diameter is A.C, the perpendicu=
lar line, which maketh a right angle with the diameter, is E.A,
whiche line falleth without the circle, and yet ioyneth ſo ex=
actly vnto it, that it is not poſſible to draw an other right line
betwene the circumference of the circle and it, whiche thyng*

is

GEOMETRICALL.

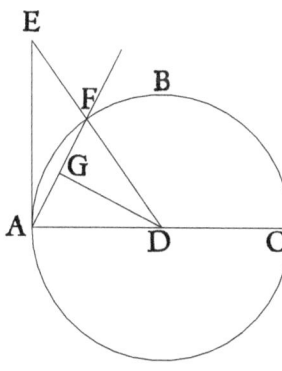

is so plainly seene of the eye, that it needeth no farther de=claracion. For euery man wil easily consent, that betwene the croked line A.F, (whiche is a parte of the circumferēce of the circle) and A.E (which is the said perpēdicular line) there can none other line bee drawen in that place where they made the angle. Nowe for the residue of the theo=reme. The angle D.A.B, which is made in the semicircle, is greater then anye sharpe angle that maye bee made of ryghte lines, and yet is it a sharpe angle also, in as much as it is lesser then a right angle, which is the angle E.A.D, and the residue of that right angle, which lieth without the circle, that is to saye, E.A.B, is lesser then any sharpe angle that can be made of right lines also. For as it was before reherfed, there canne no right line be drawen to the angle, betwene the circumference and the right line E.A. Then muft it needes folow, that there can be made no lesser angle of ryghte lines. And againe, if ther canne be no lesser then the one, then doth it sone appear, that there canne be no greatter then the other, for they twoo doo make the whole right angle, so that if anye corner coulde bee made greater then the one parte, then shoulde the residue bee lesser then the other parte, so that other bothe partes mufte be false, or els bothe graunted to be true.

The lxii. Theoreme.

If a right line doo touche a circle, and an other right line drawen frome the centre of tge circle to the point where they touch, that

k.i.   *line*

THEOREMES

line whiche is drawenne frome the centre,
ſhall be a perpendicular line to the touch line.

Example.

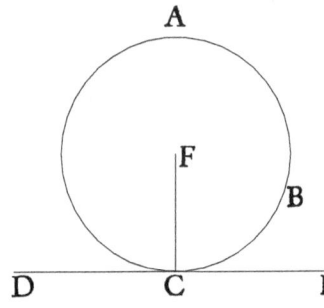

A

F

B

D   C   E

The circle is A.B.
C, *and his centre is* F.
The touche line is D.
E, *and the point wher*
they touch is C. *Now*
by reaſon that a right
line is drawen frome
the centre F. *vnto* C,
which is the point of
the touche,   therefore
ſaith the theoreme, that the ſayde line F.C, *muſte needes bee*
a perpendicular line vnto the touche line D.E.

The lxiii. Theoreme.

If a righte line doo touche a circle, and an
other right line be drawen from the pointe of
their touchinge, ſo that it doo make righte
corners with the touche line, then ſhal the cen
tre of the circle bee in that ſame line, ſo dra=
wen.

Example.

The circle is A.B.C, *and the centre of it is* G. *The touche*
line is D.C.E, *and the pointe where it toucheth, is* C. *Nowe*
it apper=

GEOMETRICALL.

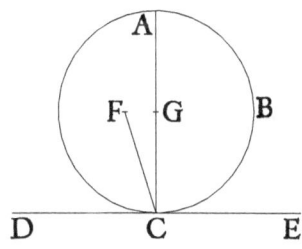

it appeareth mani=
fest, that if a righte
line be drawen from
the pointe where the
touch line doth ioine
with the circle, and
that the said lyne doo
make righte corners
with the touche line, then muste it needes go by the centre of
the circle, and then confequently it muft haue the sayde cētre
in him. For if the saide line shoulde go befide the centre, as F.
C. doth, then dothe it not make righte angles with the touche
line, which in the theoreme is fuppofed.

The lxiiii. Theoreme.

If an angle be made on the centre of a cir
cle, and an other angle made on the circumfe
rence of the fame circle, and their grounde
line be one common portion of the circumfe=
rence, then is the angle on the centre twife
fo great as the other angle on the circūferēce.

Example.

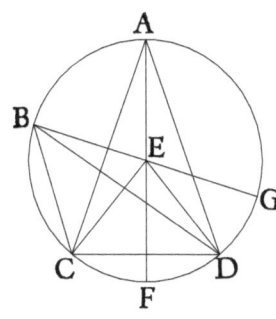

The circle is A.B.C.D, *and*
his centre is E : *the angle on*
the centre is C.E.D, *and the*
angle on the circumference
is C.A.D : *their commen*
ground line, is C.F.D, *Now*
fay I that the angle C.E.D,
whiche is on the centre, is
twife fo greate as the angle
C.A.D, which is on the cir
cumference.

k.ii.   The

THEOREMES

The lxv. Theoreme.

Thofe angles whiche be made in one cantle of a circle, muft needes be equal togither.

Example.

Before I declare this theoreme by an example, it fhall bee needefull to declare, what is to be vnderftande by the wor= des in this theoreme. For the fentence canne not be knowen, onles the very meaning of the wordes be firfte vnderftand. Therefore when it fpeaketh of angles made in one cantle of a circle, it is this to be vnderftand, that the angle mufte touch the circumference : and the lines that doo inclofe that angle, mufte be drawen to the extremittes of that line, which ma= keth the cantle of the circle. So that if any angle do not touch the circumference, or if the lines that inclofe that angle, doo not ende in the extremities of the corde line, but ende other in fome other part of the faid corde, or in the circumference, or that any one of them do fo eande, then is not that angle ac= compted to be drawen in the faid cantle of the circle. And this promifed, nowe will I cumme to the meaninge of the

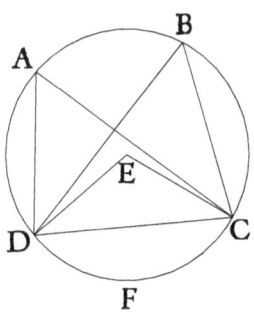

theoreme. I fette forthe a circle whiche is A.B.C.D, *and his centre* E, *in this circle I drawe a line* D.C, *whereby there ar made two cantels, a more and a leffer. The leffer is* D.F.C, *and the grea= ter is* D.A.B.C. *In this greater can tle I drawe two angles, the firfte is* D.A.C, *and the fecond is* D.B.C *which two angles by reafon they are made bothe in one cantle of a circle (that is the cantle* D.A.B. C) *therefore are they both equall.*

Now

GEOMETRICALL.

Now doth there appere an other triangle, whoſe angle ligh=
teth on the centre of the circle, and that triangle is D.E.C,
whoſe angle is double to the other angles, as is declared in the
lxiiij. Theoreme, whiche maie ſtande well enough with this
Theoreme, for it is not made in this cantle of the circle, as the
other are, by reaſon that his angle doth not light in the circum=
ference of the circle, but on the centre of it.

The .lxvi. theoreme.

Euerie figure of foure ſides, drawen in a
circle, hath his two contrarie angles equall
vnto two right angles.

Example.

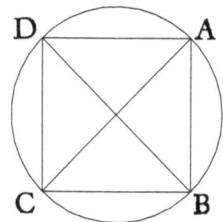

The circle is A.B.C.D, *and the*
figure of foure ſides in it, is made
of the ſides B.C, *and* C.D, *and*
D.A, *and* A.B. *Now if you take*
any two angles that be contrary,
as the angle by A, *and the angle*
by C, *I ſaie that thoſe .ij. be equall*
to .ij. right angles. Alſo if you take
the angle by B, *and the angle by* D, *whiche two are alſo con=*
trary, thoſe two angles are likewaies equall to two right an=
gles. But if any man will take the angle by A, *with the angle*
by B, *or* D, *they can not be accompted contrary, no more is*
not the angle by C. *eſtemed contrary to the angle by* B, *or yet*
to the angle by D, *for they onely be accompted* contrary an=
gles, whiche haue no one line common to them bothe. Suche is
the angle by A, *in reſpect of the angle by* C, *for there both ly=*
nes be diſtinct, where as the angle by A, *and the angle by* D,
haue one common line A.D, *and therfore can not be accomp=*
ted contrary angles, So the angle by D, *and the angle by* C,

k.iij.

haue

THEOREMES

haue D.C, *as a common line, and therfore be not contrary an=*
gles. And this maie you iudge of the residewe, by like reason.

The lxvii. Theoreme.

Vpon one right lyne there can not be made
two cantles of circles, like and vnequall, and
drawen towarde one parte.

Example.

Cantles of circles be then called like, when the angles that
are made in them be equall. But now for the Theoreme, let the
right line be A.E.C, *on whi=*
che I draw a cantle of a cir
cle, whiche is A.B.C. *Now*
saieth the Theoreme, that it
is not possible to draw an o=
ther cantle of a circle, whi=
che shall be vnequall vnto
this first cantle, that is to say,
other greatter or lesser then
it, and yet be lyke it also, that is to say, that the angle in the one
shall be equall to the angle in the other. For as in this example
you see a lesser cantle drawen also, that is A.D.C, *so if an an=*
gle were made in it, that angle would be greatter then the an=
gle made in the cantle A.B.C, *and therfore ban not they be cal=*
led lyke cantels, but and if any other cantle were made great=
ter then the first, then would the angle in it be lesser then that
in the firste, and so nother a lesser nother a greater cantle can
be made vpon one line with an other, but it will be vnlike to
it also.

The .lxviii. Theoreme.

Lyke cantelles of circles made on equall
right

GEOMETRICALL.

righte lynes, are equall together.

Example.

What is ment by like cantles you haue heard before, and it is eafie to vnderftand, that fuche figures are called equall, that be of one bygneffe, fo that the one is nother greater nother leffer then the other. And in this kinde of comparifon, they muft fo a= gree, that if the one be layed on the other, they fhall exactly a= gree in all their boundes, fo that nother fhall excede other.

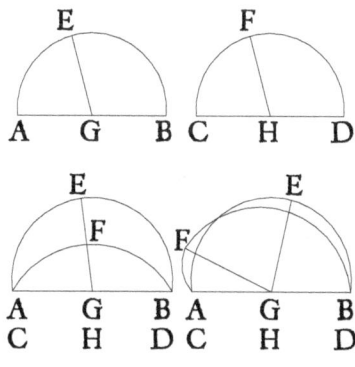

Nowe for the ex=
ample of the Theo=
reme, I haue fet for=
the diuers varieties
of cantles of circles,
amongeft which the
firft and feconde are
made vpō equall li=
nes, and ar alfo both
equall and like. The
third couple ar ioy=
ned in one, and be no

ther equall, nother like, but expreffyng an abfurde deformitee, whiche would folowe if this Theoreme wer not true. And fo in the fourth couple you maie fee, that becaufe they are not e= quall cantles, therfore can not they be like cantles, for neceffa= rily it goeth together, that all cantles of circles made vpon e= quall right lines, if they be like, they muft be equall alfo.

The lxix. Theoreme.

In equall circles, fuche angles as be equall are made vpon equall arch lines of the circum= ference, whether the angle light on the cir= cumference, or on the centre.

Example.

Firfte I haue fette for an exaumple twoo equall circles, that

is

THEOREMES

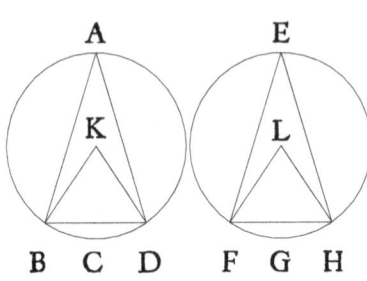

is A.B.C.D,
whoſe centre is K,
*and the ſecond cir=
cle* E.F.G.H, *and
his centre* L, *and in
eche of thē is there
made two angles,
one on the circum
ference, and the o=
ther on the centre*
*of eche circle, and they be all made on two equall arche lines,
that is* B.C.D. *the one, and* F.G.H. *the other. Now ſaieth the
Theoreme, that if the angle* B.A.D, *be equall to the angle* F.
E.H, *then are they made in equall circles, and on equall arch
lines of their circumference. Alſo if the angle* B.K.D, *be equal
to the angle* F.L.H, *then be they made on the centres of equall
circles, and on equall arche lines, ſo that you muſte compare
thoſe angles together, whiche are made both on the centres, or
both on the circumference, and maie not conferre thoſe angles,
wherof one is drawen on the circumference, and the other on
the centre. For euermore the angle on the centre in ſuche ſorte
ſhall be double to the angle on the circumference, as is declared
in the threeſcore and foure Theoreme.*

The .lxx. Theoreme.

*In equall circles, thoſe angles whiche bee
made on equall arche lynes, are euer equall to=
gether, whether they be made on the centre,
or on the circumference.*

Example.

*This Theoreme doth but conuert the ſentence of the laſt The=
oreme*

GEOMETRICALL.

oreme before, and therfore is to be vnderſtande by the ſame examples, for as that ſaith, that equall angles occupie equall archelynes, ſo this ſaith, that equal arche lines cauſeth equal angles, conſideringe all other circumſtances, as was taughte in the laſte theoreme before, ſo that this theoreme dooeth af= firming ſpeake of the equalitie of thoſe angles, of which the laſte theoreme ſpake conditionally. And where the laſte the= oreme ſpake affirmatiuely of the arche lines, this theoreme ſpeaketh conditionally of them, as thus: If the arche line B.C. D. be equall to the other arche line F.G.H, then is that angle B.A.D. equall to the other angle F.E.H. Or els thus may you declare it cauſally: Bicauſe the arche line B.C.D, is equal to the other arche line F.G.H, therefore is the angle B.K.D.e= quall to the angle F.L.H. conſideringe that they are made on the centres of equall circles. And ſo of the other angles, bi= cauſe thoſe two arche lines aforeſaid ar equal, therfore the an gle D.A.B, is equall to the angle F.E.H, for as muche as they are made on thoſe equall arche lines, and alſo on the circum= ference of equall circles And thus theſe theoremes doo one declare an other, and one verifie the other.

The lxxi. Theoreme.

In equal circles, equall right lines beinge drawen, doo cutte awaye equalle arche lines frome their circumferences, ſo that the grea= ter arche line of the one is equall to the grea= ter arche line of the other, and the leſſer to the leſſer.

l.i.   Example

THEOREMES

Example.

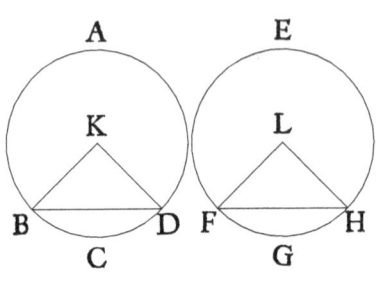

The circle A.
B.C.D, *is made*
equall to the cir=
cle E.F.G.H, *and*
the right line B.
D. *is equal to the*
righte line F.H,
wherfore it folo=
weth, that the ij.
arche lines of the
circle A.B.D,
whiche are cut from his circumference by the right line B.D,
are equall to two other arche lines of the circle E.F.H, *being*
cutte frome his circumforence, by the right line F.H. *that*
is to faye, that the arche line B.A.D, *beinge the greater arch*
line of the firfte circle, is equall to the arche line F.E.H,
beynge the greater arche line of the other circle. And fo
in like manner the leffer arche line of the firfte circle, beynge
B.C.D, *is equal to the leffer arche line of the fecond circle,*
that is F.G.H.

The lxxii. Theoreme.

In equall circles, vnder equall arche lines
the right lines that bee drawen are equall to=
gither.

Example.

This Theoreme is none other, but the conuerfion of the
lafte Theoreme beefore, and therefore needeth none other ex=
ample. For as that did declare the equalitie of the arche lines,
by the equalitie of the righte lines, fo dothe this Theoreme
de

GEOMETRICALL.

declare the equalnes of the right lines to enſue of the equal=
nes of the arche lines, and therefore declareth that right lyne
B.D, to be equal to the other right line F.H, bicauſe they both
are drawen vnder equall arche lines, that is to ſaye, the one
vnder B.A.D, and thother vnder F.E.H, and thoſe two arch
lines are eſtemed equall by the theoreme laſte before, and ſhal
be proued in the booke of proofes.

The lxxiii. Theoreme.

In euery circle, the angle that is made in the
halfe circle, is a iuſte righte angle, and the
angle that is made in a cantle greater then
the halfe circle, is leſſer thanne a righte an=
gle, but that angle that is made in a cantle,
leſſer then the halfe circle, is greatter then a
right angle. And moreouer the angle of the
greater cantle is greater then a righte angle
and the angle of the leſſer cantle is leſſer then
a right angle.

Example.

In this propoſition, it ſhal be meete to note, that there is a
greate diuerſite betwene an angle of a cantle, and an angle
made in a cantle, and alſo betwene the angle of a ſemicircle,
and y̆ angle made in a ſemicircle. Alſo it is meet to note y̆ al
angles that be made in y̆ part of a circle, ar made other in a ſe
micircle (which is the iuſte half circle) or els in a cantle of the
circle, which cantle is other greater or leſſer then the ſemi=
circle is, as in this figure annexed you maye perceaue euerye
one of the thinges ſeuerallye.

l.ij.   *Firſt*

THEOREMES

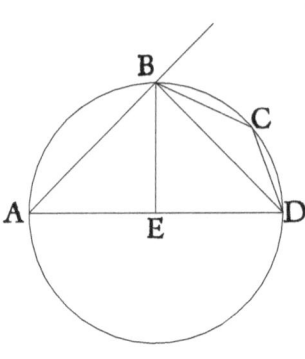

Firſte the circle is, as you ſee, A.B.C.D, and his cen= tre E, his diameter is A.D, Then is ther a line drawē from A. to B, and ſo forth vnto F, which is without the circle : and an other line alſo frome B. to D, whiche maketh two can= tles of the whole circle, The greater cantle is D.A B and the leſſer cantle is B.C.D, In whiche leſſer cantle alſo there are two lines that make an angle, the one line is B.C, and the other line is C.D. Now to ſhowe the difference of an angle in a can= tle, and an angle of a cantle, firſte for an example I take the greter cātle B.A.D, in which is but one angle made, and that is the angle by A, which is made of the line A,B, and the line A.D, And this angle is therfore called an angle in a cantle. But now the ſame cantle hathe two other angles, which be cal= led the angles of that cantle, ſo the twoo angles made of the righte line D.B, and the arche line D.A.B, are the twoo an= gles of this cantle, whereof the one is by D, and the other is by B. Wher you muſt remēbre, that the āgle by D. is made of the right line B.D, and the arche line D.A. And this angle is diui= ded by an other right line A.E.D, which in this caſe muſt be omitted as noline. Alſo the āgle by B. is made of the right line D.B, and of the arch line. B.A, & although it be deuided with ij. other right lines, of ẘ the one is the right line B.A, & tho= ther the right line B.E, yet in this caſe they ar not to be cōſide= red. And by this may you perceaue alſo which be the angles of the leſſe cantle, the firſt of thē is made of ẙ right line B.D. & of ẙ archline B.C, the ſecōd is made of the right line. D.B, & of the arch line D.C. Then ar ther ij. othr lines, ẘ deuide thoſe ij. corners, ẙ is the line B.C, & the line C.D, ẘ ij. lines do meet

in

GEOMETRICALL.

in the poynte C, *and there make an angle, whiche is called an angle made in that leſſer cantle, but yet is not any angle of that cantle. And ſo haue you heard the difference betweene an an= gle in a cantle, and an angle of a cantle. And in lyke ſorte ſhall you iudg of the āgle made in a ſemicircle, whiche is diſtinĉt frō the angles of the ſemicircle. For in this figure, the angles of the ſemicircle are thoſe angles which bee by* A. *and* D, *and be made of the right line* A.D, *beeyng the diameter, and of the halfe circumference of the circle, but the angle made in the ſemicir= cle is that angle by* B, *whiche is made of the righte line* A.B, *and that other right line* B.D, *whiche as they mete in the cir= cumference, and make an angle, ſo they ende with their other extremities at the endes of the diameter. Theſe thynges pre= miſed, now ſaie I touchyng the Theoreme, that euerye angle that is made in a ſemicircle, is a right angle, and if it be made in any cātle of a circle, thē muſt it neds be other a blūt āgle, or els a ſharpe angle, and in no wiſe a righte angle. For if the cantle wherein the angle is made, be greater then the halfe circle, then is that angle a ſharpe angle. And generally the greater the cātle is, the leſſer is the angle compriſed in that cantle : and contrary waies, the leſſer any cantle is, the greater is the angle that is made in it. Wherfore it muſt nedes folowe, that the angle made in a cantle leſſe then a ſemicircle, muſt nedes be greater then a right angle. So the angle by* B, *beyng made of the right line* A. B, *and the righte line* B.D, *is a iuſte righte angle, becauſe it is made in a ſemicircle. But the angle made by* A, *which is made of the right line* A.B, *and of the right line* A.D, *is leſſer then a righte angle, and is named a ſharpe angle, for as muche as it is made in a cantle of a circle, greater then a ſemicircle. And con= trary waies, the angle by* C, *beyng made of the righte line* B.C, *and of the right line* C.D, *is greater then a right angle, and is named a blunte angle, becauſe it is made in a cantle of a circle, leſſer then a ſemicircle. But now touchyng the other angles of the cantles, I ſaie accordyng to the Theoreme, that the .ij. an= gles of the greater cantle, which are by* B. *and* D, *as is before declared, are greatter eche of them then a right angle. And the*

l.iij.   *angles*

THEOREMES

angles of the leſſer cantle, whiche are by the ſame letters B, and D, but be on the othr ſide of the corde, are leſſer eche of them then a right angle, and be therfore ſharpe corners.

The lxxiiii. Theoreme.

If a right line do touche a circle, and from the pointe where they touche, a righte lyne be drawen croſſe the circle, and deuide it, the an‍gles that the ſaied lyne dooeth make with the touch line are equall to the angles whiche are made in the cantles of the ſame circle, on the contrarie ſides of the lyne aforeſaid.

Example.

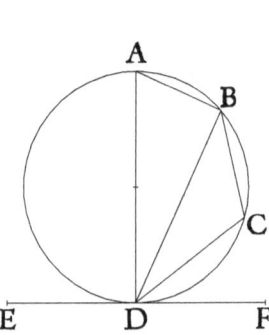

The circle is A.B.C.D, *and the touche line is* E.F. *The pointe of the touchyng is* D, *from which point I ſuppoſe the line* D.B, *to be drawen croſſe the circle, and to de‍uide it into .ij. cantles, wher‍of the greater is* B.A.D *and the leſſer is* B.C.D, *and in ech of them an angle drawen, for in the greater cantle the an‍gle is by* A, *and is made of the right lines* B.A, *and* A.D, *in the leſſer cantle the angle is by* C, *and is made of ỹ right lines* B.C, *and* C.D. *Now ſaith the Theoreme that the angle* B.D.F, *is equall to the angle made in the cantle on the other ſide of the ſaid line, that is to ſaie, in the cantle* B.A.D, *ſo that the angle* B.D.F, *is equall to the angle* B.A.D, *becauſe the an‍gle* B.D.F, *is on the one ſide of the line* B.D, *(whiche is accor‍*

dyng

GEOMETRICALL.

according to the suppostion of the Theoreme drawen crosse the cir
cle) and the angle B.A.D, is in the catle on the other side. Like=
waies the angle B.D.E, beyng on the one side of the line B.D,
must be equall to the angle B.C.D, (that is the agle by C,) whi=
che is made in the catle on the other side of the right line B.D.
The profe of all these I do reserue, as I haue often saide, to a
conuenient boke, wherein they shall be all set at large.

The .lxxv. Theoreme.

In any circle when .ij. right lines do crosse one
an other, the likeiamme that is made of the por
tions of the one line, shall be equall to the lyke=
iamme made of the partes of the other lyne.

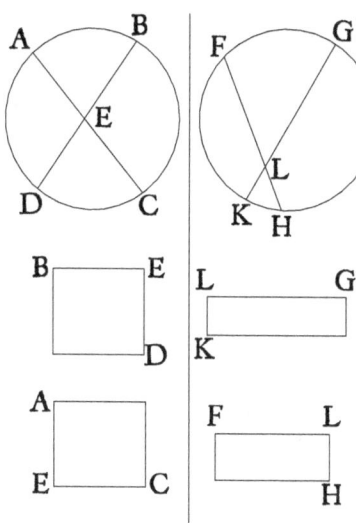

Becaufe this Theo=
reme doth serue
to many vses, and
wold be wel vn=
derstande, I haue
set forth .ij. exam=
ples of it. In the
firste, the lines by
their crossyng do
make their porti=
ons somewhat to=
ward an equalitie.
In the second the
portiõs of the ly=
nes be very far frõ
an equalitie, and
yet in bothe these
and in all other y
Theoreme is true.

In the firſt exãple the circle is A.B.C.D, in which thone line
A.C, doth crosse thother line B.D, in ẙ point E. Now if you
do make one likeiãme or lõgſquare of D.E, & E.B, being ẙ .ij.

portions

THEOREMES

portions of the line D.B, *that longfquare fhall be equall to the other longfquare made of* A.E, *and* E.C, *beyng the portions of the other line* A.C. *Lykewaies in the fecond example, the circle is* F.G.H.K, *in whiche the line* F.H, *doth croffe the o= ther line* G.K, *in the pointe* L. *Wherfore if you make a lyke= iamme or longfquare of the two partes of the line* F.H, *that is to faye, of* F.L. *and* L.H, *that longfquare will be equall to an other longfquare made of the two parts of the line* G.K. *which partes are* G.L, *and* L.K. *Thofe longfquares haue I fet foorth vnder the circles containyng their fides, that you maie fomewhat whet your own wit in practifyng this Theoreme, accordyg to the doctrine of the nineteenth conclufion.*

The .lxxvi. Theoreme.

If a pointe be marked without a circle, and from that pointe two right lines drawen to the circle, fo that the one of them doe runne croffe the circle, and the other doe touche the circle onley, the longe fquare that is made of that whole lyne whiche croffeth the circle, and the portion of it, that lyeth betwene the vtter cir= cumference of the circle and the pointe, fhall be equall to the full fquare of the other lyne, that onely toucheth the circle.

Example.

The circle is D.B.C, *and the pointe without the circle is* A, *from whiche pointe there is drawen one line croffe the circle, and that is* A.D.C, *and an other lyne is drawn from the faid*
                                                                    pricke

GEOMETRICALL.

pricke to the marge or edge of the circumference of the circle,
and doeth only touche it, that is the line A.B. And of that first
line A.D.C, you maie perceiue one part of it, whiche is A.D,

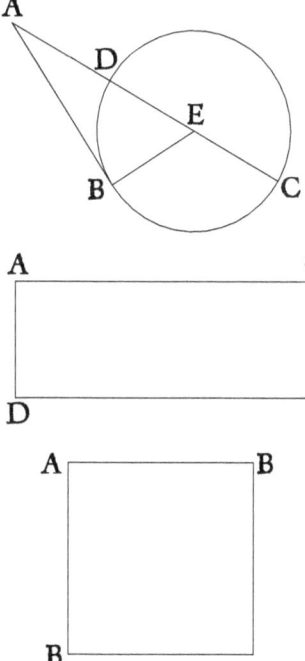

to lie without the circle,
betweene the vtter cir=
cumference of it, and the
pointe assigned, whiche
was A. Nowe concer=
nyng the meanyng of the
Theoreme, if you make a
longsquare of the whole
line A.C, and of that
parte of it that lyeth be=
twene the circumference
and the point, (whiche is
A.D,) that longesquare
shall be equall to the full
square of the touche line
A.B, accordyng not one=
ly as this figure sheweth,
but also the saied nyne=
teenth conclusion dooeth
proue, if you lyste to ex=
amyne the one by the o=
ther.

The .lxxvii. Theoreme.

If a pointe be assigned without a circle, and
from that pointe .ij. right lynes be drawen to
the circle, so that the one doe crosse the circle,
and the other dooe ende at the circumference,
and that the long square of the line which cros=

THEOREMES

ſeth the circle made with the portiō of the ſame line beyng without the circle betweene the vt= ter circumference and the pointe aſſigned, doe equally agree with the iuſte ſquare of that line that endeth at the circumference, then is that lyne ſo endyng on the circumference a touche line vnto that circle.

Example.

In as muche as this Theoreme is nothyng els but the ſentence of the laſt Theoreme before conuerted, therfore it ſhall not be nedefull to vſe any other example then the ſame, for as in that other Theoreme becauſe the one line is a touche lyne, therfore it maketh a ſquare iuſt equal with the longſquare made of that whole line, which croſſeth the circle, and his portion liyng without the ſame circle. So ſaith this Theoreme : that if the iuſt ſquare of the line that endeth on the circumference, be equall to that longſquare whiche is made as for his longer ſides of the whole line, which commeth from the point aſſigned, and croſ= ſeth the circle, and for his other ſhorter ſides is made of the por tion of the ſame line, liyng betwene the circumference of the circle and the pointe aſſigned, then is that line whiche endeth on the circumference a right touche line, that is to ſaie, yf the full ſquare of the right line A.B, be equall to the longſquare made of the whole line A.C, as one of his lines, and of his por= tion A.D, as his other line, then muſt it nedes be, that the lyne A.B, is a right touche lyne vnto the circle D.B.C. And thus for this tyme I make an ende of the Theoremes.

FINIS.

IMPRINTED at London in Poules
churcheyarde, at the figne of the Bra=
fen ferpent, by Reynold Wolfe.

Cum priuilegio ad imprimen=
dum folum.

ANNO DOMINI M.D.L.I.